The Schneider Trophy
Air Races

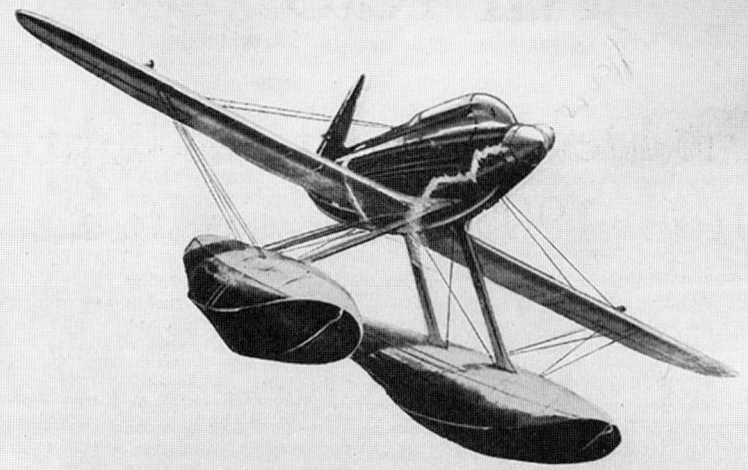

A post-1931 advert for Supermarine featuring the world speed records achieved by the S.6B.

The Schneider Trophy Air Races

The Development of Flight from 1909 to the Spitfire

Jerry Murland

Pen & Sword
AVIATION

First published in Great Britain in 2021 by
Pen & Sword Aviation
An imprint of
Pen & Sword Books Ltd
Yorkshire – Philadelphia

Copyright © Jerry Murland 2021

ISBN 978 1 52677 001 1

The right of Jerry Murland to be identified as Author of this work has been asserted by him in accordance with the Copyright, Designs and Patents Act 1988.

A CIP catalogue record for this book is
available from the British Library.

All rights reserved. No part of this book may be reproduced or transmitted in any form or by any means, electronic or mechanical including photocopying, recording or by any information storage and retrieval system, without permission from the Publisher in writing.

Typeset by Mac Style
Printed and bound in the UK by CPI Group (UK) Ltd, Croydon, CR0 4YY

Pen & Sword Books Limited incorporates the imprints of Atlas, Archaeology, Aviation, Discovery, Family History, Fiction, History, Maritime, Military, Military Classics, Politics, Select, Transport,
True Crime, Air World, Frontline Publishing, Leo Cooper, Remember When, Seaforth Publishing, The Praetorian Press, Wharncliffe
Local History, Wharncliffe Transport, Wharncliffe True Crime
and White Owl.

For a complete list of Pen & Sword titles please contact

PEN & SWORD BOOKS LIMITED
47 Church Street, Barnsley, South Yorkshire, S70 2AS, England
E-mail: enquiries@pen-and-sword.co.uk
Website: www.pen-and-sword.co.uk

Or

PEN AND SWORD BOOKS
1950 Lawrence Rd, Havertown, PA 19083, USA
E-mail: Uspen-and-sword@casematepublishers.com
Website: www.penandswordbooks.com

Dedication

To my father, 'Hugh' Howard Fergusson Murland,
who flew the Supermarine Spitfire
from 1943 to 1945.

D Flight, 74 Squadron, at Schijndel in Holland
Left to right and standing in front of a Spitfire XVI fitted with 500lb bombs are Johnnie Bennett, Geoff Lambert, Griff Griffin, Hugh Murland and Laurie Turner.

Contents

Acknowledgements viii
Introduction ix

Chapter 1	From Blériot to Prévost at Monaco	1
Chapter 2	The Second Schneider Trophy at Monaco	33
Chapter 3	The First World War	44
Chapter 4	The Post-War Years	65
Chapter 5	The Schneider Trophy 1919–1922	88
Chapter 6	The Americans Join the Fray: 1923–1926	100
Chapter 7	British Domination: 1927–1929	127
Chapter 8	The Final Flourish: 1931	155
Chapter 9	A Star is Born	167

Appendix: Schneider Trophy Results 181
Select Bibliography 183
Index 184

Acknowledgements

The author gratefully acknowledges the help and support received from Louise McIntyre, the Publishing Manager of the Haynes Group Ltd for permission to use material from *Schneider Trophy to Spitfire* and Andrew Johnston, the Managing Director of Quiller Publishing Ltd for permission to quote from Jeffrey Quill's book, *Birth of a Legend: The Spitfire*. Pen and Sword Books have also very kindly given permission to quote from Stella Pixton's book on her father, *Howard Pixton: Test Pilot and Aviator*. While I have made every effort to trace the copyright-holders of the material used, some are almost impossible to find. To that end I crave the indulgence of literary executors or copyright-holders where these efforts have so far failed and would encourage them to contact me through the publisher so any error can be rectified.

Photographs have been sourced from The National Archives, *Flight Magazine* and *The Aeroplane* as well as from my own private collection. If there are any omissions please contact the publisher who will include a credit in any subsequent editions. Occasionally some of the early photographs I have used to illustrate the text are of less than good quality, but bearing in mind popular photography developed from the Kodak box in 1890, providing of course the subject remained completely still, it is no surprise that some of these images are poor! However, I do apologize for any images that are below par. I am particularly indebted to the assistance received from the staff at the Royal Air Force Museum and the Royal Aeronautical Society, especially in the sourcing of photographs and material. A book such as this cannot be written without the co-operation of my wife Joan, who, not entirely *au fait* with the contents of the book, has nodded perceptively when I have discussed aspects of flight with her and made insightful suggestions with regard to where further information may be obtained.

Introduction

'Marine aviation has a special place in the history of flying. In spite of the difficulties encountered by its pioneers, it soon became a sporting activity which had evident military possibilities.'

Stéphane Nicolaou, writing in *Flying Boats and Seaplanes*.

The first chapter in this book is devoted to some of the early flights made from 1909 to 1913. Obviously not all those adventures, for that is what they were, are listed but perhaps some would not have been made at all if it had not been for the efforts of two brothers in Dayton, Ohio in 1903. When the bicycle-building Wright brothers took up aeronautics as a sport in 1896, they did not expect that they would become pioneers in the science of flight seven years later in 1903. Legend has it that the brothers tossed a coin to see who would be the first to fly the Wright Flyer over the sands of the Kill Devil Dunes in North Carolina. Winning the toss was older brother 36-year-old Wilbur Wright, but his

The Wright Flyer on the sands of the Kill Devil Dunes in North Carolina.

Orville and Wilbur Wright seated in what was probably the Wright Flyer III.

first attempt on 14 December 1903 was unsuccessful; three days later Orville Wright managed to leave the ground for twelve seconds, touching down 120ft away. Exchanging turns at the controls on three further occasions, Wilbur finally managed a flight lasting almost one minute and covering 852ft. Four and a half years after they had first written to the Smithsonian Institute asking for books on flying, they had achieved the impossible and flown, albeit for only a short distance, in a powered heavier-than-air machine.

The Wrights commissioned their employee Charlie Taylor to build a simple 12 hp gasoline engine; a sprocket chain drive, borrowed from bicycle technology, powered the twin propellers, which were also made by hand. In order to avoid the risk of torque effects affecting the aircraft handling, one drive chain was crossed over so that the propellers rotated in opposite directions. These propellers were mounted behind the engine in a 'pusher' configuration as opposed to a 'tractor' configuration where the propeller is mounted

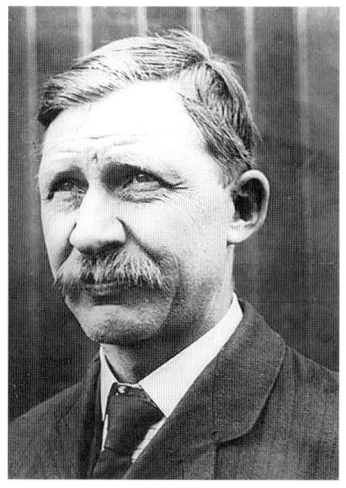

Charlie Taylor was the American inventor, mechanic and machinist who built the first aircraft engine used by the Wright brothers in the Wright Flyer. He was also a vital contributor of mechanical skills in the building and maintaining of early Wright engines and aircraft.

Jacques Schneider, standing inside the balloon basket on the left, is pictured with Maurice Bienaimé at the St Cloud Aero Parc in 1913. Bienaimé established the 1912 Gordon Bennett Balloon Race record with René Rumpelmayer when they flew 1,361 miles.

in front of the engine, pulling the machine through the air. To counter the stalling effect of the machine when turning and re-establishing lateral equilibrium, they perfected a system called wing warping, consisting of a system of pulleys and cables to twist the trailing edge of the wings in opposite directions. Although powered flight did not really come of age until 1909, the Wright brothers produced their Flyer III in 1905 which is often regarded as the first practical powered aircraft in history. It could bank, turn and fly a figure of eight with incredible ease and, what's more, could remain in the air for ninety minutes.

The concept of air racing was probably born when the 29-year-old Jacques Schneider, who was at the time the French Under Minister for Air, met the Wright brothers at Le Mans in August 1908. The early air meetings were often publicized as races, but it was not a race as we have come to know it with all the competitors starting at the same time from a common starting-point. To avoid the crowding, particularly at the corners of closed-circuit races, competitors started at intervals with the pilots racing against time rather than against each other. The concept of the fastest aeroplane and the skill of the pilot essentially took the place of reckless and foolhardy flying. Thus, the publicized 'race' to be the first pilot to cross the English Channel was in fact not a race at all, but a contest between Hubert Latham, Louis

Hubert Latham.

Louis Blériot.

Blériot and the Russian aviator, Comte Charles de Lambert. Not devoid of its own drama, Latham took off from Cap Blanc-Nez, near Sangatte, on 19 July 1909 but after only 8 miles his Antoinette IV suffered engine failure and he had to ditch in the Channel, thereby, by default, performing the world's first landing of an aircraft on the sea. Meanwhile, Louis Blériot had set up camp just under 2 miles away at Les Baraques and announced his intention to go for the prize in his Blériot XI monoplane. At about 3.00 am on 25 July, Blériot's team noticed a break in the weather, awakened him, prepared the aircraft and waited for dawn to make the attempt. Blériot took off precisely at dawn to make the first successful crossing of the English Channel by aeroplane. Of the attempt by Comte Charles de Lambert, it was reported that he had damaged both his Wright Flyers in practice and withdrew. Latham died in mysterious circumstances either during an expedition to the Congo in June 1912 or, as some sources state, he died from his injuries in America after an encounter with a buffalo.

The Russian aviator, Comte Charles de Lambert.

Prior to the meeting with the Wright Brothers, Schneider's taste for adventure took him to racing hydroplane boats and ballooning. The heir to an enormous industrial empire based on metallurgy and armament manufacturing, he became a balloon pilot with the *Aéro Club de France* in 1908 and established a high-altitude record in ballooning at 33,074ft. Schneider's thinking undoubtedly took in the inherent risk of engine failure, which was naturally greater in land-based racing aircraft, concluding that it was difficult to find an area in which a forced landing could be safely carried out. With a seaplane flying over water, a forced landing could be made in comparative safety at almost anywhere along the course. It was for this principal reason that the Gordon-Bennett races for land-based aircraft were discontinued after 1920.

On 5 December 1912, at the fourth James Gordon Bennett banquet hosted by the *Aéro Club de France*, Schneider proposed an annual contest for seaplanes, the *Coupe d'Aviation Maritime Jacques Schneider*, or the Schneider Trophy as it became more widely known, although the unofficial name, Flying Flirt, was widely used in the 1920s owing to the propensity of the trophy to temporarily lodge with each of its suitors. Schneider intended his trophy to support the technical progress of civil aviation and, as such, participants had to fly a distance of at least 150 miles over open sea, with the winner hosting the races during the next year. If a nation won the cup three times within five years, the cup would belong to them, so ending the competition. The number of laps and the distance changed from year to year, generally increasing as the quality of the planes increased, and placing were based on the highest average speed for the course. Before each machine could begin its laps, it had to remain stationary in the water for six hours and then taxi for a set distance while still in contact with the water. After the machine finally took off, it had to complete the course with any excess water that it had accumulated over that time. From time to time these pre-race conditions were changed and additional requirements were imposed by the organizing committee.

The trophy itself is made of silver and bronze and mounted on a marble base. A nude winged figure

The Schneider Trophy stands in the Science Museum in London.

representing speed is depicted kissing a zephyr on a breaking wave. The heads of two other zephyrs are also visible in the breaking wave along with the Roman god of the sea, Neptune. Today the trophy stands in the Science Museum in London. When Schneider proposed the contest it seemed most likely that the oceans would be the spawning ground of long-distance aviation; however, despite the trans-ocean commercial routes being established with flying boats, the efficiency and endurance of the landplane finally eclipsed the seaplane. His emphasis was on speed because he rightly believed that the search for speed also produced reliability and the quest for higher speed ultimately produces progress in design and equipment. For those reasons the Schneider Trophy quickly assumed the mantle of the *Blue Riband* event of the air and developed into a gigantic battle towards the attainment of high speed. As far as evidence is concerned, one only has to note that the winning speed in the trophy contests increased on eleven occasions, rising progressively over eighteen years from the 45.75 mph attained by Maurice Prévost in 1913 to the 340.08 mph of the Mitchell-designed Supermarine S.6B flown by Flight Lieutenant John Boothman during the final 1931 contest.

Jacques Schneider could have had little idea of the character that the contest would ultimately assume and probably imagined that the event would in due course be won by a Frenchman; certainly by the date of his untimely death in 1928 he would not have realized the huge effect his ideas would have on aviation and the advancement of the aeroplane. In fact, it would be quite reasonable to say that the development of flight from 1914 was, in many ways, linked to the Schneider Trophy races. Though he did not live to see the flowering of maritime aviation in the 1930s, he was able to attend the 1927 Trophy in Venice before his premature death from an appendicitis operation. The contests may not have brought the world closer to a future where all countries with access to water were connected by a network of seaplanes and flying boats, but they did evolve into an intense contest of national pride and air supremacy, one that resulted in the creation of the fastest seaplanes the world has ever seen, to this day.

Just as the Curtiss racers set the standard for the military aircraft of the 1920s, so Reginald Mitchell's designs for the contest set the standard for the late 1920s and 1930s, the knowledge gained by British designers and engine manufacturers, such as Rolls-Royce, would prove invaluable for the significant developments in aviation leading up to the Second World War. The Schneider Trophy contests ultimately set the scene for a successful generation of British fighter aircraft and, had Britain not won the Battle of Britain in 1940, it is likely that Germany would have invaded Britain and history would be drastically rewritten. Central to this success was the Rolls-Royce R engine which developed into the Rolls-Royce Merlin engine, giving the Hawker Hurricane and Spitfire the edge over their opponents.

Introduction xv

An advertisement for spectators proposing to view the 1931 contest from the sea front at Southsea.

While it is probably a little too much to lay the complete abandonment of the German Operation SEA LION at Schneider's door, it was Mitchell's design of the Supermarine Spitfire that provided the backbone of RAF Fighter Command until the arrival of the jet engine. By the war's end in 1945, the Spitfire had been modified on more than twenty occasions, and when production finally stopped in 1947, more than 22,000 Spitfires had been produced.

Even so, the first Schneider Trophy Race was clouded by farce. Out of the four planes that entered, Maurice Prévost, the French pilot awarded the victory, completed the whole course but taxied over the finish line instead of flying over it, which resulted in disqualification. One of the other contestants' machines bounced so hard on the waves and threw up so much water that the plane slowly came to a stop with a waterlogged engine. Another managed to take off 10ft into the air before the plane dived into the water and sank, which left just one more pilot, whose oil line finally burst, forcing him to land. Prévost then agreed to re-fly the final lap and was given first place by default. It was obvious from this first race that the reliability of seaplanes could be improved upon, which was very much what Schneider had intended, and great improvements were made by the second race held in 1914, again in Monaco. The first-place speed achieved in 1913 was almost doubled in this contest and also saw, for the first time, a British winner by the name of Howard Pixton.

The Schneider Trophy was put on hold during the years of the First World War and by 1918, not only had the world changed considerably but aviation had progressed sharply. Until 1923 though, Schneider's vision of practicality was still upheld and the flying boats entered by the Italians dominated the competition. In fact, in 1922 the Italians came very close to three consecutive wins and on this occasion the Supermarine Sea Lion III, England's only entry, achieved a narrow victory over the Italians, ensuring the competition would live on. In 1923, the very nature of the race was changed forever. Up until this point, the competitors of the race had been private individuals and manufacturers, and their aircraft were mostly improvements and modifications of existing First World War designs, but the Americans changed everything when they were represented by both the United States army and navy. In America, the military was deep in the disarmament of the 1918 armistice and funding was being cut quickly. To combat this, both the army and navy turned their attention to the Schneider Trophy, resulting in the development of high-quality planes designed specifically for racing: enter the Curtiss racers. Being able to recruit from the pool of government and military aircrew meant that America was represented by highly-organized, trained and experienced pilots. The result was instant. The navy took both first and second places with two Curtiss CR-3 seaplanes, averaging some 15 mph above the third-place English competitor. For years, Europeans looked down with some contempt upon American aviation, but the 1923 contest shattered this opinion and demonstrated to all that America was now back at

Introduction xvii

A Curtiss CR-3 at the 1923 Schneider Trophy contest.

the forefront of aviation technology. From this point on the Schneider Trophy became the jousting ground for the pilots and aircraft of America and Europe. One important spin-off from the advent of the Curtiss racers was the utilization by Richard Fairey of the CR-3 engine. Appreciating its full implications, Fairey lost little time in acquiring the engine and propeller and the licence to produce the engine in Britain. Around this he built the Fairey Fox which was 50 mph faster than its predecessor, the Fairey Fawn. The Fox made its maiden flight on the morning

The Fairey Fox was a direct spin-off from the Curtiss era and made its first flight on 3 January 1925.

of 3 January 1925 and performed better than advertised. Allocated to 12 Squadron, the Fox was able to penetrate any mock battle defences with almost embarrassing ease. By the time of the German invasion in 1940, the Fox was still being flown by the Belgian Air Force, but after fighting valiantly against the invader they were completely outclassed by the Luftwaffe.

The inclusion of government-sponsored racing teams saw the last vestiges of Jacques Schneider's original intentions for the Schneider Trophy contest vanish beneath a wave of nationalism. Commercial practicality was no longer a concern and if the trophy was to be retained by a particular nation, speed would be the deciding factor. This shift of intent changed the nature of the Schneider Trophy Races: they were no longer a sporting competition specifically to foster technology; they were now an iconic standard of national pride and international rivalry that would require all the resources a national budget could muster if they wanted to win. The retention of the trophy became, in effect, the world's most sought-after air-racing prize.

In 1924 the competition crossed the Atlantic to Baltimore, but in light of the entirely new level of competition, both Italy and France were unable to prepare a strong entry in time and withdrew, and despite preparing aircraft to compete, France would never again be a contender for the trophy. The English team had a promising entry with their Gloster II, but just five weeks before the race it crashed and sank, leaving the Americans with no competition. Surprisingly, America decided to postpone the race until the next year so that their rivals could compete; a grand gesture that extended the life of the contest and certainly prevented America from earning their three consecutive wins.

The American pilot Jimmy Doolittle, after completing the first Santiago to La Paz, Bolivia flight in a Curtiss Hawk, a distance of 18,000 miles crossing the Andes mountain range which rises to 15,000ft. The flight was made on 3 September 1926.

The 1925 contest was again won by America, earning them their second consecutive win in a Curtiss RC3-2, an improved version of the 1923 winner. Two days later, James 'Jimmy' Doolittle and the RC3-2 went on to break the world speed record for seaplanes, averaging 245.7 mph. As far as the aviation industry was concerned it looked very much as though America would claim the Schneider Trophy in 1926. Behind the scenes there had been talks of officially making the Schneider Trophy a biannual event to support the rising complexity and cost of developing the aircraft, a decision on which Britain and Italy were banking. However, when America announced in January 1926 that it would not be postponing the race for a year as expected, Italy and Britain were caught completely off guard with the British entry forced to drop out, leaving the Americans and Italy to do battle together. At this point Benito Mussolini entered the arena and sponsored the Italian entry with orders to both Macchi and Fiat to produce a winning entry. In a surprise underdog victory, Italian pilot Mario de Bernardi won the contest with the Macchi M.39. Italy's victory would prove to be its last, but at least it did persuade the American Congress to pull out of funding any further attempts.

Until this point, the English government had only given minimal support to their racing team, but in 1927 the government finally responded to public demand and the trophy was given the full backing of the Air Ministry and Treasury with the establishment of the Royal Air Force High Speed Flight team. For its remaining three years, the race would almost become a contest directly between the two major engineers of the time: Supermarine's Reginald Mitchell and Macchi's Mario Castoldi. In Venice in 1927 Flight Lieutenant Sidney 'Pebbler' Webster reclaimed the trophy for Britain in the Mitchell-designed Supermarine S.5, averaging over 280 mph. It was finally agreed in 1928 that the race should be held every two years in order for the entrants to prepare properly, and the extra year's grace certainly gave Mitchell the opportunity to engage the services of Henry Royce to build a more powerful engine. Castoldi's Macchis had proved fast but fragile in testing, and the Italians feared the worst. They

Flight Lieutenant Sidney Webster, winner of the 1927 Schneider Trophy contest.

were right and Flight Lieutenant Richard Waghorn achieved 328.63 mph over the 217-mile course to claim another victory for Britain. By now the Schneider Trophy had become a source of huge national pride and, despite Mussolini's anger at Italy's poor showing and throwing even more resources at winning the contest, they still lost the final competition. However, it was again touch and go for the British. Unbelievably, with Italy's government backing their attempt, the British government did just the opposite and withdrew funding for 1931. Into the breach stepped Lady Lucy Houston with a gift of £100,000 to the Treasury. Neither the French nor the Italians, blighted by more accidents, one of them fatal, turned up in 1931 and Britain won unopposed, with Flight Lieutenant John Boothman flying the Supermarine S.6B to victory. Many observers felt that Britain had shown its unsporting side in allowing the 1931 contest to proceed, but the backstage grumblings were forgotten when George Stainforth took the aircraft with its Rolls-Royce engine through the 400 mph barrier.

The Schneider Trophy's effect on military development precedes the First World War and there is an element of truth in the notion that the contest did little more than just fuel military advancements. With regard to the fighter aircraft of the First World War, the influence of the Schneider Trophy contests can be seen in the design of the Sopwith Tabloid biplane, the winner of the 1914 contest. The

The Supermarine 6.B, flown by Flight Lieutenant John Boothman, won the 1931 Schneider Trophy. The aircraft can be seen today at the Science Museum in South Kensington, London.

The Sopwith Camel was probably the most successful Allied fighter of the First World War. Later variants were fitted with a more powerful Gnome Monosoupape Type N rotary engine capable of producing 160hp. Though considered difficult to fly, the Camel owed its extreme manoeuvrability to the strong gyroscopic effect of the engine and the fact that 90 per cent of its weight was within 7ft of the nose of the aircraft.

machine had a simple wooden construction that Sopwith would later use in a long line of very successful fighters, most notably the Sopwith Camel and Snipe. The Camel is considered to have been the most successful allied aircraft of the war and it was in a Sopwith Camel that pilot Roy Brown was credited with shooting down Manfred von Richthofen in April 1918.

However, power alone would not have produced such high speed without a parallel advance in the principles of drag reduction. Here, the Schneider Trophy contests may have made their least recognized contribution to aviation. The earliest machines presented a blunt and often square frontal design, but by 1931 aircraft had been transformed beyond recognition into a streamlined version that cut through the air with accomplished ease. The fact that aircraft, impeded by rather ungainly floats, could have attained speeds of more than 400 mph post-1931 is sufficient acknowledgement of the Schneider Trophy's role in the development of aerodynamics.

The Schneider Trophy path was often extraordinary, and in encouraging the talents of such men as Mario Castoldi, Reginald Mitchell and Henry Rolls, the competition may not have remained true to Jacques Schneider's conception. Of the four principal nations that entered the contests, France got the least out of the competition but America changed the character of the race, administering the shock that stimulated the rapid advances made in both Britain and Italy. However, after

America's withdrawal from the trophy in 1926, it began to fall behind Europe in the development of engines and airframes for high-speed fighters. Of the four nations, the Italians gave the most in their determination to win the *Coppa Schneider*. They submitted entries for more races than anyone else, their designs were often the most imaginative and they lost the most pilots in preparing for the contests. Mario Castoldi had a hand in designing fighters for the Second World War but seemed to run into the same problems he encountered in the Schneider Trophy contests. He finally gave up on Fiat and had to turn to German engines. Arguably the best of his designs was the MC.202 Folgore fitted with a Daimler-Benz DB 601 engine, an aircraft that in North Africa in 1942 performed better than the Messerschmitt Bf 109.

The British made the most of the Schneider Trophy contests with Henry Folland designing the Gloster Gladiator, although by the outbreak of war its biplane format was outdated, but it still held its own against more advanced designs and defended the island of Malta from invasion. More importantly, Mitchell's work on the low-wing monoplane produced the Spitfire and the Rolls-Royce R engine fathered the illustrious Merlin, powering not only the Spitfire but also the Hurricane, Lancaster, Mosquito and, as convincing evidence of the effect of America's withdrawal in 1926, the P-51 Mustang.

The Macchi MC.202 Folgore was designed by Mario Castoldi and considered to be one of the best Second World War fighters to serve in large numbers with the Italian *Regia Aeronautica*.

Chapter One

From Blériot to Prévost at Monaco

'The collective identity of the aeroplane as revealed at Reims showed the world's new vehicle as capable of conveying two men through the air in comparative safety at some 40 mph.'

Charles Gibbs-Smith writing in *Aviation* magazine.

In 1909 the aeroplane came of age in a year marked by Louis Blériot's successful Channel crossing in July and a month later during the first great air meeting on the Betheny Plain outside Reims: *La Grande Semaine d'Aviation de la Champagne.* The meeting at Reims reinforced the message of Blériot's Channel crossing by bringing together the finest pilots and machines and showing the variety and possibilities of the new air vehicle to governments and public alike. Local French vintners and city officials founded the Reims Air Meet when they agreed to raise prize money and sponsor the air show and because of Blériot's feat the previous month, the organizers of the Reims Meet were expecting a large turnout at their event. To accommodate the anticipated crowds, they transformed Betheny into a massive aerodrome and a mini-city, building barber and beauty shops, telephone and

Louis Blériot was the first pilot to fly across the English Channel in his heavier-than-air Blériot XI monoplane on 25 July 1909, landing heavily at Northfall Meadow, Dover. The Blériot Memorial marks the spot where he landed.

2   The Schneider Trophy Air Races

Hubert Latham's
Antoinette IV at Reims.

Henri Farman at Reims.

Eugène Lefebvre at Reims sitting in a Wright biplane. He was killed later in the year at Juvisy-sur-Orge and is considered to be the first victim of an air crash.

telegraph offices, and a huge grandstand, complete with a 600-seat restaurant that overlooked the airfield. To keep people entertained between flights, they hired stilt-walkers and tightrope artists to perform.

Twenty-two aviators came to Reims to compete. All of them, save two, were Frenchmen. Some of the better-known French pilots included Blériot, Hubert Latham, Henri Farman and Eugène Lefebvre, while George Cockburn, a Scot, and Glenn Curtiss, an American, were the only foreign competitors. Although most of the pilots were experienced, there were also a few who were not. One was Emile Ruchonnet, a Swiss engineer who worked at the time for the Antoinette Company in building the Antoinette aeroplanes flown by Latham and his cousin, René Labouchère. Ruchonnet had flown his first aeroplane only two days before the meet and was tragically killed in January 1912 when his machine crashed at Senlis in France.

Glenn Curtiss at the controls of what is thought to be the Reims Racer. At Reims he was one of only two foreign competitors along with George Cockburn.

The Gordon Bennett Cup Race, or speed contest as it was commonly known, was the most important event of the Reims Air Meet and the race that everyone wanted to watch. Pilots had to fly two circuit laps consisting of 6 miles with each team allowed three entrants. Despite its comparatively meagre prize money, most spectators considered it the premier event of the week. American James Gordon Bennett, the famous publisher of the *New York Herald* newspaper and a long-time fan and sponsor of various speed contests, lent his name to the race by putting up the prize money and offering a trophy. Blériot, Europe's most celebrated aviator, was favoured to win, but as Glenn Curtiss, the lone American and a celebrated pilot in his own right made clear, the Frenchman was going to have to fight if he wanted to win the race.

By Saturday, 28 August, the day of the speed race, the field of competitors had narrowed considerably due to several crashes. At one point during the week, there had been at least a dozen disabled or wrecked planes on the field. Curtiss, fearing just such a disaster, had refused to enter his aeroplane, called the Reims Racer, in any other contests besides the Gordon Bennett Cup. All told, because of the high attrition rate, only five pilots stood ready to race for the cup: Curtiss, Blériot, Latham, Lefebvre and Cockburn. Curtiss was the first to fly the two laps around the course, averaging 46.5 mph and establishing the benchmark time of fifteen minutes and fifty seconds. Then Latham, Lefebvre and Cockburn each tried to beat his mark, but failed in the attempt; it was now up to Blériot, the last person to fly. On his first lap, Blériot led Curtiss by four seconds and was averaging 47.75 mph, but on his second lap he slowed considerably and when he crossed the finish line, Curtiss had won the race by six seconds. At first the French crowd was stunned by Blériot's performance, but after the initial shock they eventually started cheering for Curtiss and proclaimed him the new

James Gordon Bennett.

Probably the first aerobatic pilot was Eugène Lefebvre who thrilled the crowds at Reims. Lefebvre had only learned to fly that summer.

Léon Delagrange about to land in a Blériot XI at Reims. Note the number of spectators present.

'Champion Aviator of the World'. That evening also saw Henri Farman flying his Farman III and Eugène Lefebvre in his Wright Type A contesting the *Prix de Passengers*, the passenger-carrying competition. Both pilots successfully managed to take off with one passenger, but Farman won by taking off a second time with two passengers and completing a circuit. Sadly, Lefebvre was killed in an air accident at Juvisy-sur-Orge in September 1909, only nine days after the end of the Reims event.

Between 300,000 and 500,000 spectators witnessed the races and contests during the week. Of the thirty-eight aircraft originally registered to compete, only twenty-three took to the air and, of those, fifteen were biplanes and eight were monoplanes. In all, the pilots completed eighty-seven flights during the competition. For most of the people who attended the event, the Reims Air Meet showed that aviation competitions were a tremendously exciting form of entertainment and as one spectator, David Lloyd George, a future British prime minister noted, the meet also proved that 'flying machines are no longer toys and dreams...they are an established fact.' For those who had any doubts about the future of aviation, the Reims Air Meeting not only legitimized the importance and significance of flight, but also set the standard by which people would measure all future air meets.

Although the first flight in Germany was made by the Danish inventor Jacob Ellehammer on 28 June 1908, it was nearly two years later that Hans Grade won the 40,000 Reichmark prize for the first figure of eight to be flown in an all-German machine. A month later in November 1909, the Austrian Igo Etrich made the first successful flights in his monoplane at Weiner-Neustadt, a design that eventually evolved into the successful Taube in 1910 which was first produced by Rumpler. In June and July 1911 the Circuit of Germany began at Johannisthal airfield near Berlin with seventeen stages ending in Berlin. The pilots completed one stage each day, although they had more than one day's stay at some stage locations to compete in flight competitions, such as the Kiel flight meeting. The stage winner was the one who had flown through the route fastest; the overall winner was the pilot who had flown the most kilometres at the end of the advertised total route. First place went to Benno König flying an Albatros-Farman, Hans Vollmöller came second

Hans Grade seated in the Grade monoplane after winning the 40,000 Reichmark prize for the first figure of eight to be flown in an all-German machine.

A replica Rumpler-Taube built by Art Williams which can be seen at the Museum of Flight in Seattle.

and Bruno Büchner third. On 19 June, Hellmuth Hirth also set a new altitude record during a flight from Kiel at 7,217ft and later in 1911 he won the Munich to Berlin Race in a Rumpler-Taube. Flight in Germany had come of age but there was always a cost: Benno König died of injuries sustained in an air accident on 1 July 1912.

Three years before the Reims Meet, the *Daily Mail* had offered a prize of £10,000 to the first pilot to fly from London to Manchester within twenty-four hours. Only two landings were allowed and the aircraft had to start and finish the flight within 5 miles of the London and Manchester offices. At the time the prize was offered, the newspaper's money appeared to be quite safe and was believed to be unwinnable, prompting *Punch* magazine to offer a similar sum for the first man to swim the Atlantic Ocean!

Benno König, pictured on the right, won the Deutsche Rundflug in 1911 flying an Albatros-Farman biplane.

However, by 1910 the extraordinary pace of aerial development meant that a flight between London and Manchester was now possible. Present at the Reims meeting was a 30-year-old Englishman called Claude Grahame-White who was running a successful car dealership in Mayfair. Less than a year before he had scarcely bothered with aviation, but then Blériot flew across the Channel and at once Grahame-White

Claude Grahame-White.

was interested. A month later he met Blériot, and in a sudden burst of enthusiasm agreed to buy one of the famous Frenchman's latest machines. He was apparently told by Blériot that he had designed a newer and faster machine than that which had flown the Channel. It was a two-seater monoplane, powered by a 50hp Gnome rotary engine and it was entered for the Rheims Air Meet. Grahame-White said he would buy the machine at the close of the meeting. Like many other machines, the Blériot XII never saw the end of the display. It caught fire in the air and nearly burned its creator and pilot to death. In those days, however, a crash, no matter how serious, was looked upon as part of the business and in no way deterred aviators or enthusiasts.

Blériot was flying again within a few days, and Grahame-White was as keen as ever to acquire one of his machines. The monoplane was finished in November 1909, and Grahame-White immediately learned to fly it. After only half an hour's ground instruction Grahame-White flew his first solo flight. From that point in time events progressed at a startling pace.

Grahame-White's announcement that he intended to compete for and win the London to Manchester prize came as a complete surprise. He had heard that Louis Paulhan was also to compete and was determined to anticipate the French pilot in his quest for the prize. Given that his Blériot monoplane would not stand up to the rigorous demands of the race, Grahame-White bought a Farman III complete with the new 50hp Gnome rotary engine.

Henri Farman was an Anglo-French aviator and aircraft designer who was making a name for himself and his brother, Maurice Farman. Making his first flight in a Voisin-Farman biplane in September 1907, Henri Farman went on to set numerous official records for both distance and duration. These included the first to fly a complete circuit of 1 kilometre on 13 January 1908 and the world's first cross-country flight between Bouy and Reims in October. In 1909, he opened a flying school at Châlons-sur-Marne at which George Cockburn, who represented Great Britain in the *La Grande Semaine d'Aviation de la Champagne* at Reims, was the first pupil.

Henri Farman made several record-breaking flights, including one of 112 miles in three hours and another of just over 144 miles in four hours, seventeen minutes.

Grahame-White, having surveyed the route of the race between London and Manchester, decided to follow the route of the London and North Western railway and persuaded the railway company to whitewash its sleepers for 100 yards to the north of each junction. Pitted against Grahame-White's eagerness was Paulhan's experience. Paulhan was a gifted pilot and had flown with great distinction all over France and, despite crashing at Reims in 1909, had established new records for altitude and duration. In 1910 he took four machines to America where he established a new altitude record of 4,164ft and gave the newspaper baron William Randolph Hearst his first flight. It was during this

Louis Paulhan was the winner of the *Daily Mail* £10,000 prize for the flight from London to Manchester.

time at the first American Air Meeting in Los Angeles that he promised to take the young William Boeing on his first flight, but subsequently let him down in what many historians recognize as one of Paulhan's greatest missed opportunities.

Grahame-White, on the other hand, had been flying for less than six months and was a relative novice compared to his French counterpart. Nevertheless, with the advance in aeroplane and engine design, it looked as though it was only a matter of time before one or the other would manage to fly the distance. In what must have been a compliment to Henri Farman, both men opted to use the Farman biplane, although Paulhan's machine had two rudders at the tail and a shortened lower wing to give the aircraft greater speed. Grahame-White's first attempt on 23 April 1910 took him over Harrow and through the Chiltern Hills and on to Northamptonshire. After some two hours in the air he approached his first scheduled landing at Hillmorton near Rugby in Warwickshire. Breaking an undercarriage strut during the landing, he had unwittingly set a new British cross-country record during the 75 miles of his first leg. A desperately cold Grahame-White set off again at 8.15 on 24 April, but high winds forced him to land at Hademore near Lichfield. Here his valiant attempt came to an end as his Farman was blown over onto its back and irreparably damaged.

Grahame-White's attempt had roused Louis Paulhan into a frenzy of activity and he immediately set off for England. A week before Paulhan had established a world record for the longest flight of 130 miles between Orléans and Châlons, a feat that must have given Grahame-White some concerns. At 5.20 on 27 April Paulhan, intent on stealing a march on his rival, took off without a preliminary test flight and a full half-hour elapsed before Grahame-White was awakened and told that Paulhan had already taken off. As darkness fell, Paulhan was already

The Henri-Farman biplane in which Claude Grahame-White attempted to make the first flight from London to Manchester in 1910. The photograph shows the aeroplane overturned by the wind after Grahame-White had been forced down by bad weather.

almost 60 miles ahead of his rival. Even though Grahame-White's night flight from Northampton to Lichfield was designed to overhaul Paulhan, the weather was very much against him and his unreliable engine had let him down for a second time. As he landed near Lichfield he was just in time to see Paulhan taking off on his final leg of the race. At 5.25 on 28 April 1910 Paulhan landed in Manchester, cold but victorious. He had flown from London to Manchester in twelve hours, four hours and twelve minutes of which had been spent in the air.

Paulhan's flight was certified by the Royal Aero Club on 30 April and he drove immediately afterwards to the Savoy Hotel in the Strand where he was presented with his cheque. The lunch, which hosted a number of dignitaries such as the pioneer aviators Charles Stewart Rolls and John Moore-Brabazon, was also the occasion when the *Daily Mail* announced another race, this time around Great Britain. Six months later, in October 1910, Grahame-White won the second Gordon Bennett Trophy for Britain at Long Island, USA.

The year of 1910 saw the death of 23-year-old Jorge Chávez, the Peruvian aviator who was killed in September while attempting to cross the Alps in his Blériot XI. Two weeks previously he had set a new altitude record of 8,840ft. His flight was in response to the $20,000 offered by the *Aero Club d'Italia* for the first aviator to make the trip alive, and it was nearly successful. His crash on landing at Domodossola airfield saw him dying of his injuries four days later. A death that was probably given more prominence in England was that of Charles Rolls of Rolls-Royce fame who was killed in July 1910. Only a few weeks before his accident at Bournemouth he had flown from Dover to Sangatte and back without landing, making him the first man to fly the double Channel crossing. His great friend John Moore-Brabazon, who became the first resident Englishman to fly in England on 2 May 1909, was reportedly so upset by this accident that he virtually gave up flying. Moore-Brabazon was the man who, as a joke to prove that pigs could fly, had put a small pig in a waste-paper basket tied

Jorge Chávez, the Peruvian aviator who was killed in September 1910 while attempting to cross the Alps in his Blériot XI.

On 2 June 1910 Charles Rolls, pictured in the centre of the photograph, accomplished the first two-way non-stop crossing of the English Channel. Almost six weeks later, on 12 July 1910, he lost his life due to a controlling wire breaking in his Wright flyer.

to a wing-strut of his machine. This may have been the first live cargo flight by aeroplane!

At Bordeaux another pioneering aviator, Frenchman Léon Delagrange, was killed when his aircraft crashed on 4 January 1910. In 1907 he was one of the first to order an aircraft from Gabriel Voisin of the Voisin brothers, enabling them to get established as manufacturers of aeroplanes. The aircraft was the first example of what was to become one of the most successful early French aircraft, the Voisin 1907 biplane. Delagrange's first really sensational flight was witnessed on 26 October 1909 at Doncaster. Here he set a world record by flying a Blériot monoplane for 6 miles at more than 50 mph.

Possibly of more importance was Henri Fabre's first powered flight in *l'Essor* from the middle of the Étang de Berre near Marseille. The first flight from water on 28 March 1910 very nearly did not take place, and had it not been for Fabre's visit to the August Reims meeting where he was persuaded to purchase two of the

Léon Delagrange at the controls of what is thought to be the Voisin Delagrange No. 3 biplane.

On 28 March 1910 Henri Fabre flew the first seaplane named *Le Canard* at *Étang de Berre*, west of Marseille. The photograph shows the Fabre Hydroavian at Monaco in April 1911.

Seguin brothers' more powerful 50hp Gnome seven-cylinder rotary engines, it may not have taken place at all. One engine was fitted to Fabre's machine and the other was sold at a profit. The story of his success is recorded in his own words:

With my hand on the throttle lever, I let the machine accelerate away; one of the rear floats rises, so I slow down and by adjusting the neutral point of the warp I am able to alter the relative inclination of the two wings. I accelerate again, and this time both rear floats rise up. I am in the air, perfectly stable, whether gliding across this oily sea or buzzing along a dozen feet above it in the still atmosphere, the impression is the same. Throttling back, I soon see the front float gently pressing down on the water, leaving a fine trace, like that of a diamond on a sheet of glass.

It is possible that this was the flight that inspired Jacques Schneider to promote the growth of seaplanes through the Schneider Trophy contests, although there is no official evidence of their meeting. In America, Glenn Curtiss, who had met Henri Fabre at the 1910 Reims meeting, was another pioneer who was interested in marine aviation. He developed Fabre's ideas to the extent of adding a larger central float to his pusher airframe and a smaller float at each wingtip; the hydrofoil in front of the pilot's seat was to provide additional lift and protect the pilot from water spray. On 26 January 1911 the machine took off from North Island, San Diego Bay in what was the first practical seaplane flight in history and on 25 February Curtiss took off with the first passenger to fly in a seaplane. He went on to design the Flying Fish, a flying boat with a stepped hull which flew in July 1912. However, he was two months behind François Denhaut, the chief pilot at Pierre Levasseur's flying school. Denhaut's flying boat hull was shaped rather like a triangle with the point at the top and was powered by a Gnome 50hp engine. A fixed undercarriage made up of two wheels allowed the machine to take off from the ground. On 12 March 1912 the flying boat crashed while attempting a landing on the River Seine and it was not until almost a month later that Denhaut took off from Juvisy airfield but landed so close to the Seine on muddy ground that the undercarriage broke. Not to be deterred, he launched into the water, taking off and landing on a

Glenn Curtiss at the controls of his hydroplane. The photograph was probably taken at North Island, San Diego Bay.

number of occasions before returning to the bank. The prototype of a long line of flying boats had flown!

The first British attempt was left to Commander Oliver Schwann, who bought an Avro Type D and fitted floats to it. Despite not having qualified as a pilot, Schwann managed to fly it off the water on 18 November 1911 and although Schwann crashed the aircraft, this was the first aircraft take-off by a British pilot from water. Schwann's unqualified status left Lieutenant Sydney Sippe to officially fly the aircraft, this time far more successfully, on 2 April 1912, making the occasion the first *official* seaplane take-off from British waters.

Commander Oliver Schwann.

The first long-distance flights took place in 1911 with Pierre Prier flying non-stop between London and Paris in April. The Paris to Madrid race, organized by the French newspaper *Le Petit Parisien* on 21 May, was won by Jules Védrines but marked by a fatal crash involving Louis Train as the competitors were taking off. The first competitor to get away was Conneau, followed by Roland Garros and Eugène Gilbert. André Frey took off at 5.35, made a circuit of the field and landed; after some adjustments he tried again at 6.00, but damaged a wheel and had to delay his attempt for repairs. As the next competitor was not ready, Garnier took off but only made a short flight. He was followed by Jules Védrines, who immediately

Lieutenant Sydney Sippe.

The Paris to Madrid race was marked by a fatal crash involving Louis Train as he was taking off. Train is pictured here standing in front of his monoplane.

aborted his flight as his aircraft was not handling properly but crashed on landing, escaping injury but severely damaging his aircraft. At 6.22 Le Lasseur de Ranzay departed and at 6.30 Louis Train was called to the starting-line:

> As soon as I left the ground, I perceived that the motor was not working well. I was about to land, after making a turn to one side, when I saw a detachment of cuirassiers crossing the flying track. I then tried to make a short curve to avoid them, and to land in the opposite direction, but my motor at that moment failed more and more, and I was unable to undertake the curve. I raised the machine, so as to get over the troops and to land beyond them. At that very moment a group of persons, who had been hidden from my view by the cuirassiers, scattered before me in every direction. I tried to do the impossible, risking the life of my passenger to prolong my flight, and to get beyond the last persons of the group. I was about to come to land, when the apparatus, which had been raised almost vertically, dropped heavily to the ground. I got out from under the machine, with my passenger, believing that I had avoided any accident. It was only then that I learned the terrible misfortune.

The terrible misfortune described by Louis Train left the prime minister of France, Ernest Monis, unconscious and with a broken leg. Monis's son and the tycoon and aviation patron Henri Deutsch de la Meurthe were both injured. The French Minister of War, Henri Maurice Berteaux, lost an arm and sustained a fatal head wound. The crash caused a panic in the crowd, causing more injuries and the suspension of all further activity. With the approval of the injured Monis, the event continued the next day.

Jean Conneau won the Paris to Rome race in a Blériot XI. His pseudonym of André Beaumont came about because as a serving naval officer he was unable to use his real name. In the same year he also won the Round Europe Race, of which only nine pilots completed the course.

Jean Louis Conneau won the Paris to Rome race in a Blériot XI using the pseudonym André Beaumont because, as a serving member of the French armed forces, he was not permitted to use his own name. A host of sponsors including the Expo Committee of Turin, in co-operation with the Celebrations Committee in Rome, the Italian Touring Club, the Parioli Company and the Parisian *Petit Journal* offered some half a million lire in prize money, a prize that attracted twenty-six pilots and their machines. Stages were planned through Dijon, Lyon, Avignon, Nice, Genoa, Pisa, Rome, Florence, Bologna and Turin but at the Parioli aerodrome in Rome the only pilots to arrive were Jean Conneau, Roland Garros, André Frey and Renato Vidart. The race should have continued from Rome to Turin, where a large crowd, gathered for the International Aviation Week, was waiting for them. Everything was prepared, but due to bad weather only André Frey with his Morane attempted the enterprise on 13 June. The flight ended with an accident near Viterbo caused by dense fog: Frey was severely injured in the accident, being stuck for thirteen hours in the wreckage of the aircraft with a broken leg, a broken arm and a broken jaw. The hapless aviator was given a consolation prize of 10,000 lire and a gold medal for his courage and Jean Conneau was deemed the winner ahead of Roland Garros, André Frey and Renato Vidart.

The year 1911 also saw the disastrous Round Europe Race, organized by France and taking place in June and July of that year. A prize of £8,000 was offered by *Le Journal* for the entire circuit, with additional prizes for the individual stages. The course covered a little over 1,000 miles beginning in Paris and divided into stages taking in the remainder of France, Belgium, Holland and Britain with the final stages being from Calais to Dover, Shoreham and Hendon and back to Paris. Among the thirty-eight starters were a number of well-known names including Jean Conneau, Maurice Prévost, Roland Garros, Jules Védrines, Charles Weymann and James Valentine. The event was marred by three fatal accidents in the first stage: Léon Lemartin crashed on take-off from Vincennes and died on arrival at hospital; Lieutenant Pierre Princeteau crashed at Issy while attempting to land to repair a fault, his aircraft overturning and catching fire; and Pierre Landron, who was killed when his aircraft caught fire near Châteaux-Thierry. Only nine pilots managed to complete the course. In first place was Lieutenant Jean Conneau, in second was Roland Garros and Jules Védrines took third place.

After the Round Europe Race came the *Daily Mail*-sponsored Round Britain Air Race, held in July and August 1911. The prize money of £10,000 would be awarded to the pilot who could complete a 1,000-mile circuit of the British Isles in the quickest time. The rules of the race required the five stages of the course to be completed in twenty-four hours and competing machines were required to carry five official 'marks' on important components such as tail, wings and engine. Up to three of these parts could be replaced during the race, but at least two of them had to be in place when the finishing line was crossed; a rule that ensured the finishing machine was, to a large extent, the same one that began the race. Repairs were allowed, but a complete rebuild was not. The route would take the competitors from London, north to Edinburgh, then west to Glasgow, south down to Bristol and finally back to London again. The 'Circuit of Britain', as it became known, would be a test for both pilots and machines. Scheduled for the summer of 1911, thirty competitors registered their intention to take part with the Royal Aero Club.

All competitors were issued with a 22ft-long strip-map of the route, which was unwound on rollers as the race progressed. Designed by Claude Grahame-White, it showed the country on either side of the route to a depth of 7 miles, compass headings for each leg, spot heights and a cross-section of the terrain. Competitors

Roland Garros came second in the Paris to Rome race in 1911 and was the first pilot to fly across the Mediterranean Sea on 23 September 1913 in a Morane-Saulnier monoplane.

From Blériot to Prévost at Monaco   19

Charles Weymann was born in Port-au-Prince, Haiti and held dual nationalities in the United States and France. He is pictured here standing next to his Nieuport monoplane.

Jules Védrines won the first stage of the Round Britain Air Race in July 1911 but ultimately came second to Jean Conneau.

were to depart after 4.00 on 22 July in order of their race number at four-minute intervals with Jean Conneau in his Blériot XI leaving first. The 31-year-old Conneau, who had been awarded his pilot's licence in 1910, was an excellent pilot and tipped as a likely winner of the race. Among the other competitors was Samuel Cody, an anglicized American who became a British citizen in 1909 and brought a heavy Wright-based biplane named the Cathedral to the race, endearing himself to everyone with his exuberant personality. The Cathedral took off after Howard Pixton's Bristol Boxkite, but although Pixton seriously damaged his machine in a bad landing at Spofforth and effectively ruled himself out of the race, he was soon to claim fame as the first British winner of the 1914 Schneider Trophy in a Sopwith Tabloid.

Another popular entrant was the young American, Charles Weymann, a Haitian-born pilot and businessman, who was forced to return to Brooklands a few moments after taking off in his Nieuport because his map had come loose. He had to wait half an hour while the other entrants got away before he could rejoin the race, undaunted by the delay, saying rather grudgingly that he would just have to make it up later. He completed the first stage in thirteenth place.

At Hendon a large crowd of some 40,000 had assembled to see the competitors land at the end of the first stage. The first to appear was Conneau who, wasting little time, cut his engine and came straight in to his approach. On the ground, an official car whisked him away to the timekeeper's tent where his record book was written up. The entrants then began to appear regularly, but the day's winner was Jules Védrines in his Morane-Borel monoplane with a time of nineteen minutes and forty-eight seconds, a mere fifteen minutes in front of Cody. After the first day of racing the field had been reduced to seventeen entrants and day two, a long-distance stage of some 343 miles to Edinburgh, began at dawn on 24 July. Despite the fog that blanketed the ground, Jules Védrines flew over Harrogate in the lead, some four minutes ahead of Conneau. After a break at Gosforth Park, near Newcastle, Védrines and Conneau took off again for Edinburgh with Védrines arriving at 11.00 and Conneau a little over twenty minutes later. Touching down an hour later, James Valentine in his Deperdussin found himself in third position and the highest-placed British pilot. The *Harrogate Advertiser* recorded the event:

> Four minutes and 26 seconds later, a second plane landed, piloted by André Beaumont, and to the intense annoyance of the assembled dignitaries, this second flyer was also found to be French, so he, too, was immediately disqualified (*sic*). By now, the crowd was getting edgy, but when a third plane landed 24 minutes and fifty-eight seconds later, piloted by Englishman James Valentine, the cheering could be heard on West Park.

Stage three was another long-distance leg of 383 miles which terminated at Bristol with four controls marking the route: Stirling, Glasgow, Carlisle and Manchester. Poor weather and a strong westerly wind made flying a difficult business, allowing Conneau to overtake Védrines by the time he reached Stafford Park in Manchester. Védrines had, in fact, become lost near Liverpool and did not land at Stafford Park for another three-quarters of an hour after Conneau. Luck continued to desert him when he arrived at Filton, near Bristol, twenty minutes behind Conneau, having mistaken the premises of the Bristol Aeroplane Company for the official landing ground. Unless Conneau threw away his chances on the final stage to Brighton, Jules Védrines had lost the race.

The final stage was between Bristol and Brighton, a distance of 224 miles with compulsory stops at Exeter and Salisbury Plain. Conditions were much improved from the previous day and a large crowd had turned out to see the two rival Frenchmen. From the compulsory stop at Larkhill on Salisbury Plain, it was obvious that, barring mishaps, Conneau would win the race, touching down at Brooklands on 26 July. Védrines landed just over an hour later and James Valentine took third place.

The Round Britain Air Race had certainly captured the hearts and imagination of the British public and, like the Reims Air Meet, had created an awareness of what was possible. Many historians feel that the race highlighted the deficiencies of British aviation. The Bristol Aeroplane Company had entered four of its Type T machines, with none reaching Harrogate. Samuel Cody made the best flight in a British machine and was placed fifth after the British aviator Gustav Hamel, all of which gave the industry some glimmer of hope for the future. Sadly, Hamel died before reaching the age of 25, disappearing over the English Channel on 23 May 1914 while returning from Villacoublay in a new 80hp Gnome Monosoupape Morane-Saulnier monoplane.

The third Gordon Bennett Cup was held in July at the Royal Aeronautical Society's Eastchurch headquarters. The start of the competition was delayed by poor weather postponing the take-off by Gustav Hamel until 2.50 in the afternoon. Flying a Blériot XXIII monoplane which had been modified shortly before the race by having its wings clipped, he misjudged his first turn and crashed at speed, incredibly escaping without serious injury. At 3.00 Louis Chevalier took off, flying a Nieuport II which was powered by a 28hp Nieuport engine; unfortunately he was eventually forced to land after eleven laps, damaging his undercarriage in the process. Charles Weymann took off forty-five minutes later in his Nieuport, impressing spectators by the steepness of his banked turns, shortly followed by Alec Ogilvie, flying the aircraft in which he had finished third the previous year, now powered by a 50hp NEC engine. Last to take off were Édouard Nieuport and Alfred Leblanc. The final positions saw Weymann taking first place, closely followed by Leblanc and Édouard Nieuport. As one of the pre-eminent aeroplane designers and pilots of the early aviation era, Nieuport had established a new world speed record of 74.37 mph on 11 May 1911 at Mourmelon, flying his Nieuport II monoplane, powered by a 28hp engine of his own design. Later that year at Châlons,

Gustav Hamel was born in Germany and naturalized as an English citizen in 1910. He vanished over the English Channel on 23 May 1914.

he improved this time with a new record of 82.73 mph. His death, in an air accident four months after the third Gordon Bennett Cup, was a profound shock to the aviation world.

The 1912 Gordon Bennett Cup was held on 9 September in Clearing, Illinois. The race was thirty laps of a 4.14-mile course, totalling 124.8 miles. The race ended with a clean sweep for the French aviators. None of the American aircraft available to fly that day could exceed 78 mph and with Jules Védrines' practice flights in his Deperdussin monocoque averaging far better, a French win was sealed when he achieved 105.4 mph in just over sixty minutes.

The *Daily Mail* also sponsored a series of major air races, the first of which was held at Hendon Aerodrome on 8 June 1912. Known as the Aerial Derby, the competitors flew around London in a circuit of 81 miles with turning-points at Kempton Park, Esher, Purley, Purfleet, Epping and High Barnet. The winner in a field of six monoplanes and one biplane was Tommy Sopwith in a two-seater Blériot with Gustav Hamel coming in second with Miss Eleanor Trehawke Davies as passenger. In third place was William Rhodes-Moorhouse in a Radley-Moorhouse.

Two months previously Trehawke Davies had flown with Gustav Hamel on a flight from Hendon Aerodrome near London to Paris, gaining the distinction of being the first woman to fly as a passenger across the English Channel. She also flew with Hamel at the 1912 Whitsun meeting at Hendon, winning the altitude competition. She died in November 1915. Incidentally, the first woman to fly across the English Channel was the American 37-year-old Harriet Quimby, who took off on 16 April 1912 from Dover, a day after the RMS *Titanic* sank. She died tragically in July 1912 when her Blériot suddenly pitched forward during a flight

Tommy Sopwith with a Sopwith Camel outside his father's eighteenth-century Hampshire home of Compton Manor. He won the 1912 Aerial Derby in a two-seat Blériot monoplane.

William Rhodes-Moorhouse was third in the 1912 Aerial Derby. He is pictured here in the cockpit of a Blériot monoplane at Brooklands in 1912.

in the third Annual Boston Aviation Meeting, throwing Quimby and William Willard to their deaths.

The first Monaco competition for seaplanes took place in March 1912 at the behest of Georges Prade, the secretary of the Monaco Committee. Eight competitors entered with six different machines and although the first day was marked by Maurice Collieux crashing his Canard-Voisin, it was dominated by Jules Fischer in his Henri Farman biplane, who went on to win. In second place was Eugène Renaux flying another Farman, and coming in third was Louis Paulhan in his Paulhan-Curtiss Triad. Monaco had shown that, despite the excellent weather conditions, seaplanes had proved to be quite proficient. The Farnham flown by Renaux, for example, had completed more than 200 flights without any mechanical difficulties and even negotiated an 8ft swell without any problem.

In April 1912 the Geneva Seaplane Competition attracted a huge crowd who had come to see, among others, Frank Barra in his Paulhan-Curtiss, the Swiss aviator René Grandjean and Maurice Tétard in a Sommer biplane. Despite the wind and rain of the next day, each pilot made two flights.

The Paulhan-Curtiss Triad. The photograph was probably taken at Juan-les-Pins in June 1911.

Loosely based on the Voisin and Farman designs, the Sommer biplane was another successful aircraft.

Meanwhile, on 10 May in America, Glenn Martin made an impressive flight in his self-built pusher seaplane from Newport Bay in California to Avion on Catalina Island. Repairing his aircraft at Avion, he then made the return trip intact. In doing so he set two world records: the longest seaplane flight and the longest round-trip flight over water.

The second large seaplane meeting of 1912 took place at St Malo in August and was organized by the Aeronautical Commission of the *Automobile Club de France*. Attracting twelve competitors including three monoplanes, nine machines eventually started and prizes were awarded for the cumulative results over the course of three days. The competition was won by René Labouret and went some way towards proving that the seaplane had become reliable enough to cross the open sea. One of the largest gatherings of seaplanes in 1912 was held at Tamise in September on the River Scheldt. Sponsored by the *Aéro-Club de Belgique*, the meeting was to establish which machine would be most suitable for use on the River Congo and its tributaries. The race attracted fifteen competitors and was won by Jean Benoît flying a Sanchez-Besa, powered by a Renault 70hp engine.

For many the motor accident in September 1912 when Charles Voisin, the younger brother of Gabriel Voisin, was killed was a sad end to the year. Also in the vehicle was Baroness Raymonde de Laroche, the first woman to receive a pilot's licence; the crash left her severely injured. Charles and Gabriel Voisin's aircraft designs were a significant part in aviation history, particularly as Henri Farman flew a Voisin pusher biplane in most of his early flights, notably so when he became the first in Europe to successfully complete a 1-kilometre closed circuit in January 1908.

The year 1913 began with the Swiss pilot Oskar Bider making the first flight over the Pyrenees from Bern to Madrid in a Blériot monoplane, during which time

The Sanchez-Besa was powered by a Renault 70hp engine and was the winner of the Tamise meeting on the River Scheldt in 1912.

he reached an altitude of 11,500ft. The first big seaplane meeting of the year took place in Monaco, beginning on 3 April and concluding on 17 April with the first Schneider Trophy contest. The Monaco meeting had been dogged by a series of catastrophes involving the Mistral which left the Astra belonging to René Labouret damaged beyond repair and the three Deperdussins of Maurice Prévost, Louis Janoir and Émile Devienne damaged as they touched down at too high a speed. The flight from Monaco to Beaulieu and back was another disaster as the heavy swell evolved into a gale, forcing all the competitors to retire and the race to be annulled. Although not accompanied by serious injury to the pilots, the competition for the International Sporting Club held on the next day was a reminder of how dangerous

Oskar Bider pictured in the cockpit of his Blériot monoplane.

Oskar Bider was the first to fly across the Alps in both directions in his Blériot XI monoplane. He waited for thirteen days in Milan for good weather conditions and then flew back, this time crossing the Lukmanier Pass and Chrüzli Pass in northern Switzerland. This photograph depicts his Blériot taking off from Bern on his successful flight.

the sport of seaplane racing could be. There is some discrepancy among the published accounts of exactly whether Louis Gaudart won the competition, but what is certain is that Gaudart decided to thrill the spectators with a display of flying in his Donnet-Denhaut racing monoplane which he called L'Artois. Once he had finally taken off and was above the water the machine reared up twice before hitting the sea nose-first almost opposite the Tir aux Pigeons and both pilot and aeroplane vanished beneath the waves. The body of Louis Gaudart was discovered on 10 May by the lighthouse-keeper of the Monaco lighthouse.

Louis Gaudart.

There is little doubt that the Schneider Trophy meeting created the most excitement among the spectators and, as there was only one foreign entry, that of Charles Weymann, it was assumed that this was largely a French affair. The Mistral that had turned the Monaco meeting into a series of disasters had blown

Maurice Prévost eventually won the first Schneider Trophy contest for France after a misunderstanding about the finish line. The photograph shows him seated in the cockpit of his Deperdussin.

itself out by the day set aside for the Schneider Trophy, and the sea assumed a calmness that was typical of the southern French coast. Together with a clear blue sky and a gentle breeze, it was a perfect day for flying. The course began opposite the Tir aux Pigeons and then headed out across the bay to Cape Martin where a sharp turn took the competitors to Pointe de la Vieille and finally back to Monaco. The 10-kilometre course was to be flown twenty-eight times by the contestants for a total of 280 kilometres and to keep the crowd informed, the lap times of each contestant were posted on a large scoreboard at the Tir aux Pigeons.

The rules for the Schneider Trophy contest highlighted the eliminating trials that would establish how the three pilots of each nation would be chosen. The list of pilots hoping to be chosen for France originally numbered eight, but the series of catastrophes of the previous week had narrowed the field down to three: Maurice Prévost flying in a 160hp Deperdussin; Gabriel Espanet in a 100hp Nieuport; and Roland Garros flying an 80hp Morane-Saulnier. Although the team had effectively

Roland Garros, pictured here on the wing of his Morane-Saulnier, was declared runner-up to Prévost in the first Schneider Trophy contest.

chosen itself, all French contestants presented themselves for the eliminating trials. The situation regarding who was going to challenge the three Frenchmen was left to Charles Weymann, the winner of the third Gordon Bennett Cup in 1911. Although he was flying a Nieuport powered by a 160hp Gnome engine, he was in fact representing America and appeared to have dual French and American nationality despite permanently residing in France.

The fastest time recorded in the trials was that of Maurice Prévost in his Deperdussin monoplane. The aircraft had an excellent pedigree having come first and second in the 1912 Gordon Bennett Cup, which established the machine as the fastest in the world. The powerplant installed in Prévost's machine was a fourteen-cylinder Gnome rotary engine, the most advanced engine of its time, but the Nieuport, especially in the skilful hands of Charles Weymann, was a worthy contender for the trophy. On the other hand, Roland Garros, although likely to be beaten for speed by the other contestants, was another skilful pilot who was always prepared to take a calculated risk in his Morane-Saulnier monoplane.

Starting at intervals in an order of take-off that was decided by lottery, the first competitor away was Maurice Prévost. His machine taxied to the line from where the first 5-kilometres had to be covered on the surface of the water and on opening the throttle, the aircraft rose smoothly into the air and headed off across the bay. The next competitor was Roland Garros, who ran into difficulties almost immediately as he tried to take off. Bouncing jerkily along the surface of the water in apparent haste to get airborne, the continual soaking finally persuaded the engine to give up. Distraught, Garros signalled for a tow back to port from where he began his repairs. The race rules were very clear on this: his time would take into account the original start and the hour that it took for the engine to dry out would be added to his overall time. Meanwhile, at 8.50 am Gabriel Espanet had got away without difficulty with Charles Weymann following him into the air shortly afterwards. Prévost by this time was forging ahead and despite his skilful handling of the machine, the speed of Weymann's Nieuport was steadily clipping seconds off the Frenchman's lead. By lap 15 the two men were almost neck and neck, so much so that they probably failed to notice the demise of Gabriel Espanet, who had retired from the race with a misfiring engine. After lap 20 Weymann led Prévost by more than three minutes and to the spectators it looked as though the race was won. Prévost's earlier start meant that he was about to complete the race and, in accordance with the rules as he understood them, he landed short of the finish line and taxied across. Weymann had almost completed his twenty-first lap, and with the spectators and officials eying the scoreboard, it was realized that he only had to complete the remainder at six minutes per lap to win the race. Waiting to greet Weymann was the president of the Aero Club of America who had come

Charles Weymann flying his Nieuport VI during the first Schneider Trophy contest. He retired with four laps to go with a ruptured oil pipe.

to Monaco with the Stars and Stripes ready to drape over Weymann's Nieuport, but it was not to be. Suddenly, with four laps left, Weymann was seen to land; the minutes ticked by before he announced that a burst oil pipe had damaged the engine beyond repair and he was out of the race.

However, the drama was set to continue. Prévost's failure to observe the rules correctly – he should have flown, not taxied across the start line – resulted in his disqualification and to qualify in second place at all he had to return to the circuit and fly across the finish line. In typically French fashion he refused, stating it was Weymann's race and he wasn't remotely interested in second place. Of course all this took place before Weymann's failure to complete the course but was enough to galvanize Roland Garros into action once again and, with his Morane-Saulnier now repaired, he took off to complete his seaworthiness trials. Those who watched his performance in the air soon saw his ability to handle his aircraft in a skilful and practical manner and wondered what might have been had Garros been flying the faster aircraft. Excitement abounded as Prévost, persuaded by Weymann's misfortune to complete his race, was seen to be taxiing out in his Deperdussin. Flying deliberately over the finishing line, he was finally declared the winner in a fortuitous but well-deserved victory for France. Those last 50 yards had added nearly an hour onto Prévost's official time and brought his average speed down from 61 mph to 45 mph. Roland Garros, in the meantime, averaged 60 mph and successfully completed the course, but the time he had lost at the beginning proved too much to make up and he was declared runner-up to Prévost.

There was a sad aside to the French victory in that the contest gave the French aviation manufacturer Armand Deperdussin his last great victory, as four months later he was arrested in Paris on a fraud charge involving a sum of over a million pounds. He was not brought to trial until 1917, and in March he was convicted of embezzlement from his company and jailed for five years. The company was taken over and renamed the *Société Pour L'Aviation et ses Dérivés* (SPAD) by Blériot Aéronautique in 1914. Deperdussin committed suicide in 1924.

Despite the misunderstanding and the assertions of farce, there was general agreement in the aviation world that the first Schneider Trophy contest had been successful, auguring much for the future. More importantly, the 1914 contest would again be held in Monaco and the French, being on home ground and with the weight of French aviation manufacturing behind them, would start as favourites.

A day or so after the Schneider Trophy contests, Gustav Hamel made the first non-stop flight from Dover to Cologne, followed by the French pilot Marcel Brindejonc des Moulinais making a tour of the European capital cities in his Morane-Saulnier. Beginning in Paris he reached St Petersburg on 18 June and crossed the Baltic to arrive in Sweden on 25 June. His route back to Paris took him to Copenhagen and The Hague before completing his journey on 2 July.

Sadly Samuel Cody, one of aviation's great pioneers, was killed in August 1913. A flamboyant showman who was often confused with Buffalo Bill Cody, his

The wreckage of Cody's floatplane at Laffan's Plain near Aldershot. The aircraft broke up in mid-air and crashed, killing Cody and his passenger, William Evans, instantly.

grave in Aldershot Military Cemetery is adjacent to a memorial to his only son, 22-year-old Samuel Franklin Leslie Cody, who joined the Royal Flying Corps and was killed in Belgium on 23 January 1917 while serving with 41 Squadron. On 29 December 1909 Cody became the first man to fly from Liverpool in an unsuccessful attempt to fly non-stop between Liverpool and Manchester. He set off from Aintree Racecourse but only nineteen minutes later he was forced to land at Valencia Farm near Eccleston Hill, St Helens, because of thick fog. The next year he won the Michelin Cup for the longest flight made in England in December 1910. In 1911 his machine was the only British aircraft to complete the *Daily Mail*'s Circuit of Great Britain Race, finishing fourth. His death came as he was test-flying his latest design, the Cody Floatplane, which broke up at 200ft. He and his passenger, the cricketer William Evans, were killed at Ball Hill, Laffan's Plain, near Farnborough. The two men, not strapped in, were thrown out of the aircraft. Howard Pixton wrote that he was terribly upset when he heard the news and took a long time to get over the shock:

> It happened a few days before the start of the Round Britain Seaplane Race which he was entering. It wasn't due to bad piloting, but due to structural weakness. He was testing his new 100hp competition machine without its floats at Laffan's Plain and flying with him was his passenger for the contest.... Without warning the wings of his machine folded up. Cody and Evans dropped out of the machine as it turned upside down and they fell helplessly through the sky to the ground without a hope of surviving. The machine crashed into trees.

The subsequent Round Britain Seaplane Race was almost a complete washout in that only the Sopwith, flown by Harry Hawker with Harry Kauper as passenger, started the 1,540-mile course. Retiring after his aircraft was damaged in an emergency landing near Dublin, Hawker had only completed about two-thirds of the course, but by way of consolation he was given a prize of £1,000 for his effort.

As the year drew to a close Lieutenant Pyotr Nesterov of the Imperial Russian Army performed what is thought to be the first loop-the-loop manoeuvre at Kiev in a Nieuport IV monoplane, to be followed a few days later in France by Adolphe Pégoud. September was an action-packed month with the indomitable Roland Garros making the first crossing of the Mediterranean from St Raphael to Bizerte in Tunisia, while Maurice Prévost won the fifth Gordon Bennett Cup in his Deperdussin. The second Aerial Derby was held on Saturday, 20 September 1913, and was flown over a slightly longer 94-mile course, alterations having been made because the original route crossed areas prohibited under the recently-passed Air

Lieutenant Pyotr Nesterov carried out what is believed to be the first loop in his Nieuport VI monoplane on 27 August 1913. He is pictured here standing next to a Nieuport VI, believed to be the machine in which he performed the first loop.

Navigation Order. As well as the *Daily Mail* trophy and £200 prize, a trophy and three prizes of £100, £70 and £25 were given by Shell for the winner of a handicap competition. First place went to Gustav Hamel flying a clipped-wing Morane-Saulnier with Harold Barnwell in second place flying a Martinsyde and Australian Harry Hawker in third flying a Sopwith. We will hear much more of Harry Hawker in subsequent chapters, but suffice to say for the moment that he was the chief test pilot for the Sopwith Aviation Company in 1913 and it was Tommy Sopwith's design of the Sopwith Tabloid that astounded the world at Monaco in 1914.

Harry Hawker was placed third in the second Aerial Derby in a Sopwith biplane.

Chapter Two

The Second Schneider Trophy at Monaco

'France, who had been so well ahead having trained 344 pilots by 1910 to our fifty, and who had claimed so many of the Daily Mail *prizes, and had done so well in the British Military Aeroplane Trials of 1912 was now, apparently, left quite stunned.'*
Howard Pixton, speaking after the 1914 Schneider Trophy contest.

The course for the 1914 event was the same one that had seen the French win the initial Schneider Trophy and was held at the conclusion of the Hydro-Aeroplane Meeting at Monaco in April 1914. On this occasion, rather than the competitors for the Hydro-Aeroplane Meet flying round a number of laps of the course, an aerial rally was proposed which took as its aim the development of tourism by air over land and sea, and was seen to be a greater test of the aviator's skill. With several optional starting-points ending in either Marseilles or Genoa, a relay was established to allow for either a change of aircraft or the addition of floats in order for the competitors to make the 130-mile sea crossing to Monaco. Thus, at the end of the overland routes, competitors had the choice of converting their aircraft for water-borne excursions, for which forty-eight hours was allowed, or use a second aircraft of the same type which was already converted with floats. Accordingly, when those pilots who were fortunate enough to complete the course came into touchdown at Monaco, each would have completed a distance of 803 miles. Roland Garros proved to be the victor of the meet, reaching Monaco in a total flying time of twelve hours and eleven minutes.

On 8 April the French held their eliminating trials to select the team that would contest the Schneider Trophy contest. Each was required to fly four laps of the 10-kilometre course and of the six French competitors only Gabriel Espanet in his Nieuport was able to complete the distance. Pierre Levasseur managed two, Roland Garros in his Morane-Saulnier managed one and Prévost, Louis Janoir and Marcel Brindejonc des Moulinais failed to complete even one and were designated reserves. Those pilots that had at least completed a lap of the circuit – Espanet, Levasseur and Garros – were nominated as the French team. Buoyed up perhaps by the victory in the 1913 Gordon Bennett Cup race where Prévost had taken first place in his Deperdussin, the French still felt confident of gaining the trophy for the second time. Nevertheless, an element of doubt was now creeping in. Without Prévost and his streamlined Deperdussin, the question going through the minds of the French team now was whether the Nieuports had enough speed to counter any challenge from other competitors.

America was once again represented by Charles Weymann in a Nieuport powered by a 160hp Le Rhône rotary engine but on this occasion he was accompanied by William Thaw II, who was the first to fly up the East River in New York under all four bridges. Thaw was originally flying a rather tired Curtiss biplane pusher flying boat that his father had bought him and appeared to be somewhat relieved at the loan of a Deperdussin. He later went on to become an ace with the *Escadrille Americaine* during the First World War.

Ernst Stoeffler entered the competition at Monaco with an Aviatik Arrow.

The German pilot Ernst Stoeffler had also entered the competition with an Aviatik Arrow powered by a 150hp Benz engine, a machine that ultimately gave birth to the Aviatik B1. Ernest Burri, a Swiss competitor, was also among the entrants flying an FBA flying boat constructed by Franco-British Aviation and powered by a 100hp Gnome Monosoupape rotary engine driving a two-bladed pusher propeller. However, it was only capable of a top speed of about 68 mph and as such presented little problem to the French. According to Pixton, also entered was a young Dutch aviator named Anthony Fokker flying a Fokker W.1. Little, if anything else, is known about him in the contest except that he crashed before the

Ernest Burri flew in a single-seat FBA Type A flying boat. The aircraft shown in the photograph is a two-seat Type B.

Anthony Fokker in his 1910 aircraft which he called *de Spin*. It was later destroyed by his business partner who flew it into a tree.

race and effectively ruled himself out. Fokker, as we know, went on to be one of the mainstay German constructors of the First World War.

This left the British entry of Irish-born Lord John Carbery in a Morane-Saulnier and Howard Pixton in an unheard-of Sopwith aircraft. Without doubt the Sopwith machine was the dark horse of the competition and had been manufactured by Tommy Sopwith, a man we had already met when he won the 1912 Aerial Derby. Qualifying for his pilot's licence in November 1910, he made numerous pioneering flights and established a flying school at Brooklands. Among his pupils at Brooklands were Major Hugh Trenchard and Harry Hawker. In 1913 he formed

The prototype Sopwith Tabloid was a two-seater and is seen here on an early test flight.

the Sopwith Aviation Company at Kingston upon Thames where he worked in partnership with Harry Hawker and Fred Sigrist; while Hawker concentrated on test flying, Sopwith and Sigrist would turn their talents to design. One of the first aircraft produced was the Sopwith Tabloid powered by an 80hp Gnome rotary engine. Initial fight tests carried out at Brooklands indicated that it was fast; Sopwith was of the opinion that no monoplane of the period would be able to match its rate of climb or manoeuvrability. As for the pilot, Sopwith engaged the services of the highly-experienced Howard Pixton to continue the test-flying programme and fly the Tabloid at Monaco while Harry Hawker was in Australia.

Well aware that success on an international stage was highly desirable, Sopwith decided that a floatplane version might stand a good chance of victory in the 1914 Schneider Trophy. He powered the aircraft with the new 100hp nine-cylinder Gnome Monosoupape engine, which he brought home with him in his luggage from the Seguin factory in Paris. The Monosoupape engine was a rotary engine design first introduced in 1913 by the Gnome Engine Company and used a clever arrangement of internal transfer ports and a single pushrod-operated exhaust valve to replace the many moving parts found on more conventional rotary engines. Sopwith's next task was to design a suitable float configuration along with his boat-builder Syd Burgoyne. Fortunately Sopwith had already built the Bat Boat, which had won the Mortimer Singer Prize in July 1913, and now with Burgoyne they concentrated their efforts into designing a central main float and two wingtip floats for the Tabloid. Initial tests on the River Hamble were something of a disaster and Pixton was thrown into the water before the machine cartwheeled and sank! Recovering the stricken aircraft, Fred Sigrist and his team worked day and night to restore the airframe while Burgoyne, looking at a more conventional flotation

The original Sopwith Bat Boat was test-flown off the water at Cowes in March 1913 and badly damaged when overturned by a storm. This photograph shows a modified and rebuilt aircraft.

The Tabloid being tested on the Thames.

arrangement, sawed the central float in half and, discarding the wingtip floats, replaced them with a tail float. This made the aircraft sit back, but it appeared to be the answer. With a matter of weeks before Monaco, the Tabloid was again tested on the Thames where they clashed with Thames Conservancy officials. However, at least this rather clandestine occasion assured Sopwith and his men that the Tabloid would pass the flotation tests. However, more than mere flotation tests was required and chief mechanic Victor Mahl in particular wanted to be assured of the Tabloid's take-off performance. Accordingly, the next morning at an ungodly hour the Tabloid was launched into the Thames again, this time at Ham, opposite Glover's Island, a section of the river that came under the jurisdiction of the Port of London Authority:

> We received permission to fly, and with time running out quickly, I made another attempt to test it. Once satisfied with its control on the water, I flew off with the engine misfiring a bit to Eel Pie Island, a little island in the middle of the Thames at Twickenham and managed to reach a good speed of 85 mph, but it was barely a 3-mile flight, just a hop in fact, there just wasn't the time to fly further.

Clearly the engine had not got over its ducking in the River Hamble, but Pixton and Mahl were sure the aircraft had a chance of success at Monaco. Apart from Harold Perrin and Harry Delacombe, the two representatives of the Royal Aero

Victor Mahl, Howard Pixton and Tommy Sopwith standing in front of a Tabloid single-seat scout.

Club, Tommy Sopwith surrounded himself with the Sopwith team which included Victor Mahl and Syd Burgoyne. Once they had got possession of the aircraft, the diminutive Tabloid was uncrated and assembled and when the French first saw the aircraft their mirth was almost uncontrollable and their delight was not confined to the machine. Pixton and his apparent inexperience convinced the French that the trophy was theirs for the taking. Fortunately the weather took a turn for the worse and the contest was postponed for twenty-four hours, giving the Sopwith team an additional day in which to ensure the Tabloid was ready for the task ahead. Even so, Mahl insisted on sleeping close by the machine to guard it, forsaking the comforts of the Hotel Bristol. Tragically he died after an operation for appendicitis in April 1915 while he was at Southampton testing seaplanes.

The aircraft was not looking its best. The engine, which was immersed in the River Hamble, was still coated in rust and running so erratically that it was worked on until late in the night before the contest. It only remained to see if the Tabloid flew correctly without any major adjustments, and that would have to wait until the contest itself. Early on the Sunday morning the Tabloid was loaded with fuel and Pixton taxied out to the mouth of the harbour, opened the throttle and was airborne after a run of about 100ft. Eight minutes later, after an impressively fast run, he landed neatly and taxied to the bottom of the slipway. There was one fault: the engine was over-revving and a coarse-pitch propeller was substituted which apparently eased the problem. The only other modification was to the fuel system, and to be on the safe side a small tank holding 6 gallons of fuel was added and connected to the main tank. Another precautionary measure was a heavier pair of stay wires spliced into the float chassis. The element of doubt that had crept

The Tabloid at Monaco. Tommy Sopwith is standing on the extreme right.

into the French team's psyche had now increased: would this little Sopwith beat all the other competitors?

Soon after Pixton's landing the breeze freshened up and the sea became decidedly choppy, a situation that proved fatal to Lord John Carbery and Ernst Stoeffler, who damaged their aircraft and put them out of the contest. For Stoeffler it was the end of his Schneider Trophy hopes but the nationalistic Carbery, no doubt aided by a further postponement of twenty-four hours, managed to persuade Louis Janoir to lend him his Deperdussin.

In 1914 the rules of the contest had eliminated the taxiing test of the previous year and substituted them with two touch-downs within a specified

Lord John Carbery. The aircraft behind him is a Morane-Saulnier monoplane.

distance, rather along the lines of the circuits and bumps associated with land-based aircraft. Monday morning dawned with a clear sky and a slight wind along with a reasonably smooth sea and the first away were the two French favourites in their Nieuports, Levasseur and Espanet. Meanwhile, Burri in his FBA flying boat thrilled the crowd on the terraces above the Tir aux Pigeons with a series of long, bounding moves which eventually launched his aircraft into the air. All three competitors completed the touch-downs correctly, although Espanet overran the allotted space, requiring a second attempt which was added to the time of his first lap.

Pixton's machine was the fourth away and, making short work of the take-off, headed towards the seaworthiness test which was accomplished with hardly any reduction in flying speed. Described as a 'neat piece of flying', he completed the first lap, cutting in close to the marker pylons:

> At each pylon I cut my turns and banked well over much to the surprise of everyone. But I had a special technique. Most pilots flew close into the pylons and then went wide, but the quickest way was to advance fairly widely then skim round. It was at once seen that the Sopwith was very much faster than any of the others and it was reported that no waterplane had ever been seen banking like that at Monaco, if anywhere.

When Pixton's time for the first lap was announced, the crowd were astonished: it had taken him a mere four minutes and twenty-seven seconds, half the best French time and although the Nieuports were being flown with great skill, it was proving too much for their twin-row rotary engines. With engine trouble forcing Carbery to withdraw in his borrowed Deperdussin, it became apparent that if the Sopwith could keep going, a British win was on the cards. Back on the ground Garros and Weymann were watching the race avidly and refusing to start until Pixton had completed the course. Then the unthinkable happened. The Monosoupape engine began misfiring on one cylinder and there was no telling what would happen next. Pixton's time on his fifteenth lap fell and by the eighteenth his lap times were down to the level of Espanet's and threatened to put him out of the race altogether. For six further laps the Monosoupape engine faltered and misfired with Pixton tearing off the pieces of paper on the dashboard that he used to count the completed laps. With all the makings of a French classical drama, the lap times of the two

Howard Pixton was an accomplished aviator and during his long career from 1910 to 1918 he flew about eighty different types of aircraft and logged more than 3,500 flying hours.

Pixton on the Schneider Trophy course at Monaco; the Tabloid is flying low over the casino.

Frenchmen began to fall away and after seventeen laps Espanet was forced down with Levasseur coming down a lap later, leaving just Pixton and Burri in the air! Then, with the drama at its height, the engine of Pixton's machine appeared to recover and the Monosoupape began to settle down on eight good cylinders:

> So far I'd been flying low, but I took the machine higher so I would be able to turn into the wind and glide down should the engine fail altogether. For the next few laps my lapping speed was irregular, but as my engine gave no further trouble, I came down lower once again. Lap after lap went by and the Tabloid kept going.... Burri was flying much slower than I was, so I passed him several times, then he lost about thirty minutes for refuelling. I still had ample fuel to finish the race, and during the last six laps I reached an excellent regularity of speed.

When Pixton crossed the finishing line he was given an average speed of 86.78 mph, beating Prévost's true average speed of the previous year by 25 mph. By prior arrangement with Tommy Sopwith, Pixton continued for another two laps of the course to achieve a new floatplane record of 86.6 mph over a measured 300 kilometres:

> The wind had risen, below me the choppy sea looked ominous and I had visions of turning over at the height of victory as I came in to land, but all went well. At the end of thirty laps, I made a good landing in the rough water.

Howard Pixton is pictured here on the port float of the Tabloid after winning the Schneider Trophy contest.

Mahl, who had been out in a motorboat all the time I was racing, ready for any eventualities, steered towards me looking as pleased as Punch and fixed a rope to the Tabloid to tow me back to the harbour.… I left the cockpit and stood on the float while under tow to add extra weight to the front as the floats became half-submerged and the tail dragged in the water when it was not moving under its own power.

Jacques Schneider on the right congratulating Howard Pixton (in the flying helmet) on winning the contest.

At this point Garros and Weymann accepted defeat and the race was practically over apart from Burri's marathon attempt to complete the course and the belatedly unsuccessful attempts by both Prévost and Levasseur to try to beat Pixton's time in order to salvage some French prestige. Eventually Burri finished in second place with an average speed of 51 mph.

It was interesting that the two aircraft that completed the Schneider Trophy course were both powered by Gnome Monosoupape engines; in each case the twin-row rotary engines of the other aircraft had failed from the old curse of overheating.

Chapter Three

The First World War

'The impact of aviation development on the air war during World War 1 was profound unlike much of the warfare being waged below; the air war was constantly dynamic, not just in the physical sense, but also in terms of its evolution.'
                    Hugh Cowin, writing in *Allied Aviation of World War 1*.

The victory of Tommy Sopwith's Tabloid at Monaco in 1914 shook the French aviation establishment to its roots and focused attention onto the debate between the monoplane and biplane. The thick tapered wing that later became standard for monoplanes had yet to be developed and, at the time, the wings of the monoplane and biplane were so lightly constructed that external wire bracing was essential. This drag imposed by the wire bracing was so high that the advantages of the monoplane were almost obscured. As demonstrated by the Tabloid at Monaco, the biplane had undoubted advantages in lift, structural strength and lightness and, despite the performance of the monoplane, the tendency was now to revert to the biplane. Hitherto the monoplane had been the established favourite

The Tabloid fitted with a strengthened undercarriage served with the Royal Naval Air Service.

for speed and manoeuvrability; now the Tabloid had been produced, which could convincingly beat the monoplane.

Single-seat variants of the Tabloid went into production in 1914 and thirty-six eventually entered service with the Royal Flying Corps (RFC) and Royal Naval Air Service (RNAS). These were deployed to France at the outbreak of war where they were used as fast scouts. The Tabloid was also used as a bomber: on 22 September 1914 Tabloids mounted the first raid by British aircraft on German soil, and in their most famous mission two RNAS Tabloids flying from Antwerp on 8 October 1914 attacked the German Zeppelin sheds at Cologne and Düsseldorf. The Cologne target was not located, the railway station being bombed instead, but the Zeppelin shed at Düsseldorf was struck by two 20lb bombs dropped from 600ft and Zeppelin *Z IX* was destroyed. During 1915 attempts were made to use Tabloids to intercept Zeppelins over the North Sea, launching them from seaplane carriers including HMS *Ben-my-Chree* and *Engadine*, but these efforts were largely unsuccessful due to heavy seas either making take-off impossible or damaging the floats.

British designers naturally followed the Sopwith Tabloid lead and when the French government copied Britain's example by putting restrictions on the monoplane for her military programme, French designers were forced to turn to

Oswald Boelcke was an advocate for the biplane being adopted by the German Air Service.

Rudolf Berthold was one of the German pilots who finally came out in favour of the biplane.

The Fokker Scourge took place from August 1915 to early 1916, when the Imperial German Flying Corps equipped with Fokker Eindecker fighters gained an advantage over the Royal Flying Corps and the French *Aéronautique Militaire*. This photograph depicts a captured Fokker Eindecker with British roundels.

the biplane. German designers, influenced heavily by Oswald Boelcke, rejected the Fokker Eindecker in 1916 after the death of Max Immelmann and turned to the Halberstadt D-Series biplane. *Oberstleutnant* Wilhelm Siegert commented that

> The start of the Somme battle unfortunately coincided with the low point in the technical development of our aircraft. The unquestioned air supremacy we had enjoyed early in 1916 by virtue of our Fokker monoplane fighters had shifted over to the enemy's Nieuport, Vickers and Sopwith aircraft in March and April.

It was left to Rudolf Berthold, the German flying ace who shot down forty-four enemy aircraft, to finally state that what was needed was a small biplane that was easily manoeuvrable in combat.

In 1911 Blériot rather courageously withdrew his monoplanes from service for special strengthening, citing aerodynamic and structural difficulties, a simple act that caused doubts about the monoplane's structure. A year later the RFC was formed by Royal Warrant and came into being in May when the Air Battalion was absorbed into the military wing of the new Corps. (The RNAS was smaller in size, being formed from the naval aviators based at Eastchurch.) The monoplane certainly had its fair share of accidents and fatalities when compared to the biplane, particularly as the higher wing loading of the monoplane and the higher stalling speeds on landing made it more dangerous to fly. A case in point was the Bristol Coanda monoplanes, a series of monoplane trainers designed by the Romanian Henri Coandă for the British and Colonial Aeroplane Company. However, on 10

The Coanda Monoplane at Filton, near Bristol, in November 1912.

September 1912, one of the aircraft crashed in Oxfordshire, killing both Lieutenants Edward Hotchkiss and Claude Bettington of the RFC. While this was traced to one of the bracing wires becoming detached, the result was a temporary ban on monoplanes by the military wing of the RFC.

Another casualty of the widespread belief held by the RFC of the unsafe nature of the monoplane was the Frank Barnwell-designed Bristol M.1 Monoplane Scout that went into service in the Middle East and Balkans in 1917. A single-seat tractor monoplane, it was some 30 to 50 mph faster than the Fokker Eindecker and French Morane-Saulnier, but was believed to possess too great a landing speed to be handled by French airfields and consequently never served on the Western Front. It is interesting to note that in 1919 the Chilean military pilot, Lieutenant Dagoberto Godoy Fuentealba, flew from Santiago to Mendoza in Argentina and back in an M.1C, making the first flight across the Andes Mountains. The British ban on monoplanes was relatively short-lived but manufacturers had already begun to tool up for biplane production, making the manufacture of the monoplane almost impossible without spending vast sums. The biplane was here to stay for the foreseeable future.

Certainly had the First World War not intervened it is highly likely that the monoplane would have recovered from this setback but, in war, speed has to be balanced by load and a low landing speed that would give the pilot room to operate from restricted airfields. At the same time, a good rate of climb and a high ceiling were essential. With faster engines, and to some extent design, the biplane soon equalled and surpassed the pre-war racing monoplane, the result being that the development of the monoplane was placed on the back burner.

The front view of the Bristol M.1C monoplane.

One could argue that the First World War was the catalyst that moved the aircraft from being the plaything of wealthy sportsmen to a vital military weapon that was marked by improved performance. At the war's end the biplane still dominated the aviation world, but it is doubtful that the four years of war were responsible for a parallel rise in design and development. The DH.4, for example, a conventional tractor biplane designed by Geoffrey de Havilland, continued long after the war had concluded and apart from some engine development remained much the same and was even manufactured by the United States well into the 1930s.

Both the Allies and the Central Powers developed aviation during the First World War at an extraordinary rate. Each power sought to achieve and maintain air supremacy and, regardless of what some historians have written, each side was heavily influenced by the other.

However, at the start of the war in 1914 there was some debate over the usefulness of aircraft as a weapon of war; a worth that the aircraft quickly proved to be vital. Certainly the British establishment had considered the aeroplane as a useful reconnaissance vehicle as far back as 1911 when it became clear that the development of aircraft had reached the point where they were of military significance. France, the world leader in aviation at the time, had more than 200 aircraft in military service, in direct contrast to the 19 military aircraft possessed by Britain. The only practical step that had been taken by the War Office was the creation of an establishment for the scientific examination of the various problems involved in aircraft design. After some consultation with the Royal Aero Club and various aircraft manufacturers, they announced their *Specification for a Military Aeroplane* in late December, and trials were held in August 1912 at Larkhill at the

The First World War

With the help of his family and friends Cody built the Cody Mark V, mainly from the remains of the Mark III, in just four weeks. It was powered by the 120hp Austro-Daimler engine which was relatively unscathed from the crashes it had been in. Note the triangular rudders similar to those on the monoplane. It was this machine that won, albeit with a British engine, both the International Division and the British Division of the 1912 military trials.

Military Aeroplane Competition. Thirty-two aircraft entered and the winner was the Cody biplane, which successfully passed all the tests. As a result, the Cody biplane was purchased by the RFC, with an order placed for a second example to be built by Cody. Several more of the competitors' machines were also purchased by the RFC, including the Blériot XI-2, two Bristol Coanda monoplanes and two Gnome-powered Deperdussins. We have already heard how the crash involving the Bristol Coanda monoplane in September 1912 contributed to the British ban on monoplanes. Even so, the initial campaigns of 1914 went a long way towards the acceptance of aircraft as a tool of reconnaissance: it rapidly became clear that an aircraft could at least locate an enemy force and provide some warning as to their intentions. A case in point was the intelligence provided by the RFC on 22 August 1914. Captain Lionel Charlton and Lieutenant Vivian Wadham made the crucial observation of the German First Army's approach towards the flank of the British Expeditionary Force, a move which threatened the envelopment of the BEF. This information allowed the British Commander-in-Chief, Field Marshal Sir John French, to realign his front and prevent the annihilation of the BEF during the first weeks of the war. Aeroplanes were now becoming sufficiently reliable to play a significant part in aerial warfare, particularly in fighter, reconnaissance, target and gunnery spotting duties and, as the war encouraged the development of aircraft to operate across the battlefields of the Western Front, British aircraft manufacturers were given a considerable boost.

Despite Britain being surrounded by sea and offering generous sheltered areas for machines to take off and land, there was not a comparable development with maritime aircraft as there had been with landplanes. The relatively small number of seaplanes that were in operational use during the war were little more than slightly improved versions of aircraft that had competed in the Schneider Trophy contest at Monaco in 1914. There had been some progress with flying boats due to the long-range reconnaissance required and of these, the Felixstowe F.2A and the F.3 were probably the most significant. Based on the American Curtiss H.12, they set a design trend for the next two decades, arriving in service in the spring of 1917 just as Germany's infamous U-boat campaign was peaking. Its career closely paralleled the Short Sunderland of the 1930s and firmly established the reputation of flying boats as weapons. Incidentally, it was Winston Churchill who allegedly first coined the term 'seaplanes' while answering a question in Parliament in 1913. Consequently at the end of the war there were two categories of maritime aircraft: seaplanes that were mounted on floats, and flying boats that were designed with a boat-like hull. It is also of note that in 1917 Hubert Scott-Paine, managing director of what became the Supermarine Aviation Works Ltd, appointed the young Reginald Mitchell as his personal assistant.

The Felixstowe F.2 was a 1917 British flying boat class designed and developed by Lieutenant Commander John Porte at the naval air station at Felixstowe, adapting a larger version of the Felixstowe F.1 hull design and the Curtiss H-12 flying boat. This photograph is of a Felixstowe F.2A taken in the air from another Felixstowe flying boat.

The First World War was responsible for nearly 10 million soldiers, sailors and airmen who died in battle or as a result of battle. The Allies lost almost 6 million men, while the Central Powers lost about 4 million. A considerable number of the military deaths in the First World War were in battle, unlike the conflicts that took place in the nineteenth century when the majority of deaths were due to disease. On 4 August 1914 many of the pioneer aviators that had graced the world scene with their presence before the war had already been called up into the uniform of their respective countries, while others waited for the inevitable.

Of the French and British pioneer aviators mentioned in the text, Frenchman Roland Garros is probably the most well-known for his exploits during the First World War. In 1928 the Roland Garros tennis stadium in Paris was named in his memory and the French Open Tennis Tournament takes the name of Roland Garros from the stadium in which it is held. As a reconnaissance pilot with the *Escadrille* 26, Garros visited the Morane-Saulnier Works in December 1914 where he was introduced to Saulnier's work on metal deflector wedges attached to propeller blades, thereby allowing a machine gun to be fired through a propeller. Garros eventually had a workable installation fitted to his Morane-Saulnier Type L aircraft and achieved the first ever shooting-down of an aircraft by a fighter firing through a tractor propeller. On 1 April 1915 he shot

On 5 October 1918, on the eve of his 30th birthday, Roland Garros took part in one last fateful mission over the Ardennes, along with five other French aircraft. Four of them had left to chase a German aircraft when a squadron of six Fokker planes suddenly appeared. Garros dived in for the fight and never came back.

down two more German aircraft. Seventeen days later Garros's fuel line either clogged or his aircraft was hit by ground fire, with the result that he glided to a landing on the German side of the lines. Failing to destroy his aircraft completely before being taken prisoner, the gun and armoured propeller remained intact. It was reported that after examining the plane, German aircraft engineers, led by Anthony Fokker, designed the improved interrupter gear system. In fact the work on Fokker's system had been going on for at least six months before Garros's aircraft fell into their hands. With the advent of the interrupter gear the tables were turned on the Allies, leading to what became known as the Fokker Scourge. After almost three years in captivity Garros managed to successfully escape on 14 February 1918 together with fellow aviator Lieutenant Anselme Marchal. Rejoining his squadron, he settled into *Escadrille* 26 and claimed two victories on 2 October 1918, one of which was confirmed. Three days later, on 5 October 1918, he was shot down and killed near Vouziers, Ardennes, a month before the end of the war and one day before his 30th birthday. His adversary was probably

Anselme Marchal survived the war but died in 1921.

the German ace Hermann Habich from *Jasta* 49, flying a Fokker D.VI. Anselme Marchal died in 1921 aged only 38.

Another French pilot was Marcel Brindejonc des Moulinais. His trip from Paris to London and Brussels and returning to Paris in a Morane-Saulnier C monoplane with a double crossing of the Channel drew accolades from far and wide. He also crossed the Baltic Sea in a Morane-Saulnier during his trip, visiting all the European capitals. As an NCO pilot he flew reconnaissance missions during the Battle of the Marne in 1914 for which he was mentioned in dispatches. Promoted to lieutenant on 26 December 1915, he was awarded the Croix de Guerre six months later. After the Battle of Champagne his health became increasingly poor and he had to rest in Britain. On 28 August 1915 he became the chief pilot at the Morane-Saulnier flight school at Le Bourget where he stayed until May 1916 when he joined *Escadrille* N25. In October 1916, flying a SPAD VII, he was shot down and killed along with the French fighter ace Maxime Lenoir.

Marcel Brindejonc des Moulinais was shot down and killed on 18 August 1916 in Vadelaincourt near Verdun.

The 31-year-old Louis Paulhan was mobilized as a pilot with the rank of lieutenant on 15 September 1914 and served initially in northern France near to Amiens. Probably best known for his victory in the London to Manchester race in 1910, he was transferred to the Serbian front in 1915, where he was not only the most experienced but also the oldest in his squadron, commanding a squadron of ten Maurice Farman aeroplanes. The Serbian campaign was unsuccessful, but Paulhan is credited with the world's first medical evacuation by air when he flew the seriously ill Milan Stefanik to safety. Decorated with the Croix de Guerre, he returned to France where he was involved with the manufacture of propellers for the French military.

Louis Paulhan in the cockpit of a Maurice Farman biplane of *Escadrille* MS 99.

Jean Louis Conneau, who flew as André Beaumont, was already a serving officer when war broke out and as a flying boat pilot he commanded squadrons at Nice, Bizerte, Dunkirk and Venice. He also worked at Franco-British Aviation perfecting flying boats for the French navy from 1915 until 1919.

Apart from being the chief test pilot at the Voisin factory and instructor at the Voisin Flying School before the war, the 34-year-old Maurice Collieux was another

André Beaumont won the Circuit of Britain race in 1911. This photograph shows Beaumont landing at Brooklands in his monoplane to win the race.

Maurice Collieux photographed in 1910 at the controls of a Voisin biplane.

aging pilot who was mobilized in 1914. At the flying school he taught, among others, Louis Paulhan, Sanchez Besa and Henry Fournier to fly. Mobilization saw him assigned to Pau as chief pilot at the Voisin school and then to Ambérieux in March 1915. He was demobilized with the rank of captain and the Military Medal, having achieved more than 3,400 flying hours.

Jules Védrines in the cockpit of his Morane N. Védrines used the system developed by Roland Garros to fire a machine gun through the propeller of his plane, using a deflector placed on the blades of the propeller to deflect bullets that might have struck them. He died in April 1919.

A man who perhaps had a more exciting war was Jules Védrines. Mobilized in January 1915 with the rank of corporal into the famous Stork *Escadrille*, he was largely involved in clandestine missions, landing behind enemy lines in the Verdun area to drop or pick up agents. His aircraft was named *La Vache* and was emblazoned with a picture of a cow in homage to his family's roots in the Limousin region. On 15 July 1915 he was mentioned in the French army Order of the Day for his work with the Sixth Army, for whom he had flown more than 1,000 hours on reconnaissance missions. His death came in April 1919 during the inaugural flight of the Paris-Rome line aboard a Caudron C.23 twin-engine aircraft which crashed at Saint-Rambert-d'Albon. The accident also cruelly took the life of his mechanic Guillain.

Taking part in the Paris to Madrid Race in 1911 was the jovial Eugène Gilbert, but four years later he was in uniform as a combat pilot with *Escadrille* MS23. Along with the French pre-war pilot Adolphe Pégoud, Gilbert was one of the first pilots to become an ace, shooting down five or more enemy aircraft. On 27 June 1915 Gilbert was interned after force-landing his Morane-Saulnier fighter in Switzerland while returning home from bombing the Zeppelin sheds at Friedrichshafen. He later succeeded in escaping from captivity and returning to France. He was killed on 17 May 1918 when testing a new aircraft at Villacoublay.

Eugène Gilbert was killed on 17 May 1918 when testing a new aircraft at Villacoublay.

Claude Grahame-White pictured with his wife, formerly Miss Dorothy Taylor, sometime in 1915.

British aviators needed little encouragement to join up and among these was Claude Grahame-White, who was probably one of the most well-known pioneer aviators in Britain. He was credited with the first night flight during his failed attempt to win the London-Manchester prize in 1910, and victory in the second Gordon Bennett Trophy at Long Island, USA. Grahame-White also flew his Farman biplane over Washington DC and landed on West Executive Avenue near the White House. Rather than being arrested, Grahame-White was applauded for the feat by the newspapers. At the outbreak of war he was appointed as flight

In 1910, while in Washington DC, Claude Grahame-White landed and took off on West Executive Avenue between the White House and the State War and Navy building. This photograph shows Grahame-White taking off in his Maurice Farman biplane.

Geoffrey de Havilland (left) with Eddie Cosh and his DH 60 Moth named *Corsicanfly* at Stag Lane Aerodrome in 1933. Cosh was secretary of the Society of Model Aeroplane Engineers.

commander in the RNAS and made the first night patrol over London in search of a Zeppelin reported crossing the Essex coast. In August 1915 he was recalled to superintend the construction of aircraft for the government.

Aircraft designer Geoffrey de Havilland was commissioned as a second lieutenant in the RFC on 2 September 1912 and a reserve officer on 24 November. In December 1913 de Havilland was appointed an inspector of aircraft for the Aeronautical Inspection Directorate but, unhappy at leaving design work, he was recruited in 1914 to become the chief designer at Airco in Hendon where he designed many aircraft all designated by his initials DH. Large numbers of de Havilland-designed aircraft were used during the First World War and flown by the RFC. On 5 August 1914, he was promoted to lieutenant and briefly stationed in Montrose on the east coast of Scotland where he flew a Blériot, protecting British shipping from German U-boats. After a few weeks he was released from this duty and returned to Airco, although he nominally remained in the service until the end of the war. He is probably most well-known for his design of the de Havilland Mosquito, considered to be one of the most versatile aircraft of the Second World War and we should also not forget the de Havilland Comet, the first jet airliner to go into production.

Future politician John Moore-Brabazon, having virtually given up flying after the death of Charles Rolls, joined the RFC on the outbreak of hostilities. Moore-Brabazon was awarded the first Royal Aero Club Aviator's Certificate and was the first British resident to officially take to the skies in 1909. He served on the Western Front where he played a key role in the development of aerial photography

John Moore-Brabazon became the first resident Englishman to make an officially recognized aeroplane flight in England on 2 May 1909. He is photographed here with the famous pig which is reputed to have been the first animal to fly.

and reconnaissance. On 1 April 1918, when the RFC merged with the RNAS to form the Royal Air Force, Moore-Brabazon was appointed as a staff officer and was promoted to the substantive rank of lieutenant colonel in the RAF on 1 January 1919 in recognition of his wartime services. Relinquishing his commission that year, he finished the war with the rank of lieutenant colonel and was decorated with the Military Cross on 1 January 1917.

Another distinguished airman was George Cockburn. He represented Great Britain in the first international air race at Reims and co-founded the first aerodrome for the army at Larkhill. He also trained the first four pilots of what was to become the Fleet Air Arm. In 1913, as war approached, Cockburn was appointed to be an inspector of aeroplanes in 1914 for the Aeronautical Inspection Directorate of the RFC at Farnborough. In the 1918 New Year's Honours List, he was awarded an OBE for his services. On the other hand Alec Ogilvie, who came third in the Gordon Bennett competition at Belmont Park, New York in 1910 was commissioned almost immediately as an RNAS officer with the rank of squadron commander. Initially given responsibility for overseeing flying training at the Naval Flying School, Eastchurch, he took command of the aircraft repair depot at Dunkirk in 1916 and was promoted to acting wing commander on 31 December 1916. In early 1918 Ogilvie reported on flight tests of the Sopwith Snipe, stating that its flying qualities were bad; however, he was quite rightly overruled by Trenchard and Brooke-Popham. Incidentally, the Snipe was the brainchild of Tommy Sopwith and, in the opinion of the author, was superior in every way to the best the Germans could offer at the time. On 1 April 1918, along with all other RNAS personnel, Ogilvie transferred to the newly-established Royal Air Force in the rank of temporary lieutenant colonel. He was injured in a flying accident on 8 June 1918, resigning in 1919.

Irishman Denys Corbett-Wilson, the man who first flew from Great Britain to the

George Cockburn at the controls of his Maurice Farman biplane at Reims in 1909.

The Wright brothers built the Wright Model R especially for Alec Ogilvie. He flew it in the Gordon Bennett Air Race at Belmont Park, New York in October 1910, coming in third with an average speed of 55 mph. The photograph shows the aircraft on display at Belmont Park.

Denys Corbett-Wilson with a Blériot XI-2 at Highcliffe, near Christchurch on 18 June 1913.

island of Ireland on 22 April 1912, joined the RFC at the outbreak of war. Posted to 3 Squadron, he and his observer were on a reconnaissance mission in a Morane Parasol in May 1915 when their aircraft was struck by an enemy shell. Both were reported to have been killed instantly.

James Valentine joined the RFC in 1914. In 1911, flying a Deperdussin, he was one of only four airmen to complete the Circuit of Britain race and was the only British aviator to compete in that race. He was also third in the 1912 Aerial Derby, flying a Prier monoplane. In August 1916 Valentine was selected to head a training mission to Russia and sent on 20 October 1916 with the intention of training Russian pilots to fly British-built aircraft. Like the other British forces in Russia, Valentine became involved with the 1917 Revolution and its initial aftermath and sadly he died of wounds at Kieff in Russia (now Kiev, Ukraine).

James Valentine is commemorated on the Archangel Memorial to men who died in the North Russian campaign and whose graves are not known.

A BE2 similar to the aircraft that Rhodes-Moorhouse flew at Kortrijk. This particular aircraft is thought to have been that which was flown by Montrose-based pilot Lieutenant Harvey-Kelly.

Taking second place in the Aerial Derby of 1912 in a Radley-Moorhouse monoplane was William Barnard Rhodes-Moorhouse, the first RFC aviator to be awarded the Victoria Cross in the First World War. At the outbreak of war Rhodes-Moorhouse enlisted in the RFC and as a second lieutenant was posted to Farnborough. Seeking to serve on operations, he obtained a posting to 2 Squadron in March 1915 at Merville, flying the B.E.2. On 26 April 1915 at Kortrijk, Belgium, Rhodes-Moorhouse swept low over the railway junction that he had been ordered to attack. He released his 100lb bomb and was immediately plunged into a heavy barrage of small-arms fire from rifles and a machine gun in the belfry of Kortrijk church. He was severely wounded by a bullet in his thigh, and his aeroplane was badly hit. Returning to the Allied lines, he again ran into heavy fire from the ground and was wounded twice more. He managed to get his aircraft back, and insisted on making his report before being taken to the Casualty Clearing Station. He died the next day on 27 April 1915. His son, William Henry Rhodes-Moorhouse, flew in the Battle of Britain with 601 Squadron and was shot down and killed in September 1940.

Surviving the war was Frank McClean, who in 1914 made a flight following the course of the Nile between Alexandria and Khartoum in a specially-built four-seater aircraft, the Short S.80. In 1912 he flew his Short-Farman through Tower Bridge. On the outbreak of war in August 1914 McClean joined the RNAS and carried out patrols in the English Channel before becoming chief instructor at Eastchurch. He transferred to the Royal Air Force when it was formed in 1918, but he resigned in 1919.

Starting from the Isle of Sheppey, McClean flew his Short-Farman floatplane through Tower Bridge and continued upstream, dipping under the remaining bridges as far as Westminster.

Sydney Sippe, the RNAS pilot who made the first *official* take-off from water in Britain in 1912, took part in the first bombing raids on Düsseldorf and Cologne. After failing to find his designated target, he bombed Cologne railway station, causing serious damage. This was followed by the celebrated attack on the Zeppelin sheds and factories at Friedrichshafen on 21 November 1914, one of the first long-distance bombing missions. Sippe and two other pilots flew 125 miles from Belfort, France over mountainous terrain and in difficult weather. Reaching the target area, Sippe crossed Lake Constance in mist while under heavy fire, descending to just 10ft above the water in order to use the mist as cover. Despite their aircraft taking damage, the three pilots succeeded in bombing their targets. Although substantial damage was claimed at the time and in some later histories, the damage inflicted was slight. One pilot, Eugène Gilbert, who we have already met, was shot down and captured, but Sippe and the third pilot returned safely. Sippe received the French Legion of Honour immediately after the Friedrichshafen raid at the request of General Joffre himself. Sippe was also awarded the Distinguished Service Order in the 1915 New Year's Honours List.

John Carbery, the 10th Baron Carbery, was to become an archetypal eccentric Anglo-Irish lord, playing a little-known role in the War of Independence on the side of Irish freedom. Succeeding to the landed title in 1868 when he was only 6 years old, he is most well-known in aviation circles as a team member of the 1914 Schneider Trophy contest and for his part in various air races, including the London to Birmingham race in 1914 during which he crashed in his Bristol Scout. At the age of 14 he went to Cork secretly and bought his first car and later he acquired his own aeroplane with which he gave aeronautical exhibitions to the bewilderment of the citizens of Cork. A few years later he flew at the Clonakilty and Bandon shows. Volunteering on 19 August 1914, he joined the RNAS as a

The First World War   63

The Bristol Boxkite, similar to the one used by Howard Pixton. This photograph depicts a Bristol Boxkite in front of the British and Colonial School hangars at Larkhill.

temporary flight sub-lieutenant, bringing along his own aircraft. During the earlier part of the war he was flying over the German trenches and throwing bombs out of his cockpit, but little is known about his subsequent career in the RNAS. After the war he displayed his nationalism by publicly flying the tricolour over Castlefreke. He sold the 1,100-acre estate in 1919 for a fraction of its worth, the drop in value a result of the intensifying Tan War. In 1921 Lord Carbery rejected his last British connection by having his name changed by deed poll to plain John Evans Carbery from John Evans-Freke, 10th Lord Carbery.

Howard Pixton started his flying career in 1910 at Brooklands, gaining Royal Aero Club Aviator's Certificate No. 50 and becoming friendly with Alliott Verdon Roe (A.V. Roe). Together they started the Avro School of Flying and Pixton became the first test pilot for Avro. In 1911 he joined the Bristol Aircraft Company and spent a great deal of his time teaching pupils to fly on the Bristol Boxkite. In 1914 he joined Tommy Sopwith and won the Schneider Trophy at Monte Carlo flying the Sopwith Tabloid. The 1914 win was Britain's first international victory with a British plane, giving Howard Pixton the status of being the man who put Britain in the lead in aviation for the first time. Later that year, at the outbreak of war, he was commissioned in the RFC and joined the Air Investigation Department at the Royal Aircraft Establishment, Farnborough. He retired, aged 60, in 1918 after the conclusion of the First World War, writing that flying would never be the same again:

It had grown from child to adult overnight and a new type of people had appeared on the scene, different from the closely-knit crowd of the old days…. Aviation had advanced more than it would have done in twice or thrice as many years of peace conditions and beside the small scouts of 1914, there stood huge multi-engine machines with great changes in performance.

While his observations about the nature of flying were correct, what Pixton failed to mention was the huge effect the Sopwith Tabloid had on the subsequent manufacture of aircraft during the war. Indeed, it would be another six years before Reginald Mitchell designed the Supermarine S4 monoplane to compete in the Baltimore Schneider Trophy contest.

Chapter Four

The Post-War Years

'American Charles Lindberg grasped the $25,000 Orteig prize, first offered in 1919 for the first direct flight between New York and Paris – twice the distance of Alcock and Brown's 1919 transatlantic flight.'

Peter Almond, writing in *A Century of Flight*.

Before we continue with the story of the Schneider Trophy contests, let us briefly take a look at a few of the other aerial developments that were taking place after the First World War. Once the Armistice was declared in November 1918 there was a considerable number of aircraft surplus to requirements and at the same time there were large numbers of redundant service pilots, all of which contributed to the rise of commercial aviation. Many of the pilots that had served their countries now had the choice of returning to their previous occupations, if indeed they still existed, becoming aviation businessmen or barnstormers. Flying from town to town and showing off their skills, the barnstormers sold aeroplane rides and performed stunts in the air. Among them was Charles Lindbergh, who began flying as a barnstormer before he enlisted in the United States army in 1924 and trained as an Army Air Service Reserve pilot. Barnstorming was arguably the first form of civil aviation in that pilots would charge civilians for the privilege of being taken aloft.

Mackenzie-Grieve and Harry Hawker.

The Sopwith Atlantic biplane flown by Hawker and Mackenzie-Grieve.

One of the first flights of 1919 was the abortive attempt to cross the Atlantic Ocean by Harry Hawker and Lieutenant Commander Kenneth Mackenzie-Grieve in a Sopwith Atlantic. They were forced to ditch in the sea and were given up for lost until the Danish steamer SS *Mary* announced their rescue. Harry Hawker was killed in July 1921 when his Nieuport Goshawk, the aircraft he was to fly in the Aerial Derby, crashed in a park at Burnt Oak, Edgware, not far from Hendon Aerodrome. His loss reverberated far and wide and the papers were full of it for weeks after the event; at his inquest it was shown that he had a critically-diseased spine and a related haemorrhage that possibly occurred during his last flight.

Arthur Whitten Brown and John Alcock.

The story of the first non-stop transatlantic crossing began in 1913 when the *Daily Mail* offered a prize of £10,000 to the first aviator to accomplish the task. At the time it was considered an impossible task, particularly as it had only been ten years since the Wright brothers had managed to get airborne in North Carolina. Nevertheless, the first direct non-stop flight was made in June 1919 when pilot Captain John Alcock and navigator Lieutenant Arthur Whitten Brown made the crossing in a converted Vickers Vimy bomber. Leaving Newfoundland on 14 June, they crash-landed near Clifden in Ireland the next day. Their flight of some 1,950 miles had taken just over sixteen hours at an average speed of 118 mph. Tragically, Alcock was killed six months later while flying a new Vickers amphibious aircraft, the Vickers Viking, on his way to the first post-war aeronautical exhibition in Paris when he crashed in fog at Cottévrard near Rouen in Normandy. Brown retired into private life and died in 1948.

On 12 November 1919, Ross Smith, assisted by his brother Keith and two mechanics, Wally Shiers and Jim Bennett, set out to fly from England to Australia in a large Vickers Vimy bomber, the same type of aircraft used by Alcock and Brown to cross the Atlantic earlier in the year. The England to Australia Air Race was an initiative of the Australian government headed by Prime Minister William Hughes to promote aviation in Australia. It was an epic twenty-eight-day flight, and on their arrival, the pioneering flyers were welcomed home as national heroes. Their £10,000 prize money was shared equally, the two brothers were knighted and the mechanics commissioned. The next proposal, to fly round the world in a Vickers Viking amphibian, ended in disaster. Ross and his long-serving crew member Jim Bennett were test-flying the aircraft at Weybridge near London in

Ross and Keith Smith with their mechanics Jim Bennett and Wally Shiers standing in front of their Vickers Vimy.

The Farman F.60 Goliath was initially designed in 1918 and with the new passenger cabin arrangement the aircraft could carry up to fourteen passengers.

April 1922 when it spun into the ground from 1,000ft, killing both men. The flight was abandoned and the bodies of Sir Ross Smith and Lieutenant Bennett were brought home to Australia. After a state funeral Smith was buried in Adelaide on 15 June 1922.

France actively encouraged the *Lignes Aériennes Latécoère* to open its service from Toulouse to Barcelona in December 1918, but it is commonly accepted that the first civilian airline for passengers was probably operated by the German *Deutsche Luft-Reederei* in February 1919 between Berlin, Leipzig and Weimar. In the same month the French Farman Company opened a service between Paris and London in converted Goliath bombers. The German airline prompted Louis Blériot, Henri Farman and others to establish the *Compagnie des Messageries Aériennes* (CMA) in April 1919, a commercial mail and freight service using ex-military Bréguet 14s and flying between Paris and Lille.

In Britain the first daily commercial flight opened on 24 August 1919 flying from London to Paris with a fleet of Airco DH.4As. The Aircraft Transport and Travel Company soon gained a reputation for reliability, prompting the loan of six RAF Airco DH.9A aircraft to operate the airmail service between Hawkinge and Cologne. In Britain and France, and to an extent America, the first transports were adaptations of wartime bomber aircraft which themselves were biplanes, but by 1922

The DH.34 could carry up to ten passengers.

the British airlines commissioned specially-built single-engine aircraft such as the DH.34 which could carry up to ten passengers. Larger transports, soon to be called airliners, were adapted twin- or four-engine bombers such as the Vickers Vimy or the Farman Goliath. A number of other airlines were opened in 1919 including the Dutch *Koninklijke Luchtvaart Maatschappij* (KLM), who established a route to the Netherlands East Indies by way of Dum Dum outside Calcutta. In its early days, KLM used Dutch-built aircraft such as the four-seat Fokker F.2 and the five-

The Fokker F.2 was one of KLM's early aircraft and could seat four passengers.

passenger version for its flights. British Imperial Airways was also formed in 1924 and principally served the British Empire routes to South Africa, India and the Far East, including Australia, Malaya and Hong Kong. Imperial Airways never achieved the levels of technological innovation of its competitors and was merged into the British Overseas Airways Corporation (BOAC) in 1939.

America was comparatively slow to open airlines and it was not until 1927 that a passenger route was opened along one of the various contract mail routes which had been in operation since 1918. The Colonial Air Transport Company opened the first passenger service between New York and Boston in April and the scheduled Transcontinental Air Mail service, flown between New York and San Francisco, began on 8 September 1920. The route was laid out in July and August by the First World War flying ace Eddie Rickenbacker and Bert Acosta, who had helped fly the first experimental through flight carrying about 100 letters which landed at Durant Field. The transcontinental mails were originally flown only during daylight hours while being entrained at night, although on 22 February 1921 a night-time leg on this route (Omaha to Chicago) was flown for the first time with James 'Jack' Knight as the pilot. The first daily Transcontinental Air Mail service involving both day and night flying over the entire route was opened on 1 July 1924, which reduced the time of the trip from more than seventy hours to a schedule of just over thirty-four hours westbound and thirty-two hours eastbound. In addition to New York and

Bert Acosta won the 1921 Pulitzer Trophy Race flying a Curtiss R-1.

Captain Edward Vernon Rickenbacker leaning against his SPAD S XIII in 1918.

San Francisco, the route included thirteen intermediate stops where mails were exchanged and aircrew relieved.

Bertrand 'Bert' Acosta was known as the bad boy of the air. He first rose to prominence as the winner of the 1921 Pulitzer Trophy Race flying a Curtiss Navy racer, setting a new world record for closed course racing of 176.7 mph. In April 1927 Acosta and Clarence Chamberlin set an endurance record of fifty-one hours, eleven minutes in the air. He later commanded the Yankee Squadron during the Spanish Civil War. During his colourful career he received numerous fines and suspensions for flying stunts such as flying under the Whittemore Memorial Bridge in Naugatuck. He died in 1951.

On 5 October 1922, Lieutenants John Macready and Oakley Kelly set a world endurance record of thirty-five hours, eighteen minutes and in May 1923 both men made the first non-stop coast-to-coast flight from Long Island, New York to San Diego, California in just under twenty-seven hours. En route, they made the first in-flight aircraft engine repair replacing a defective voltage regulator switch while the single-engine, high-wing Fokker T-2 Liberty monoplane continued its journey westward. The flight also set a new distance record for a single cross-country flight.

In May 1926 Commander Richard Byrd, a naval veteran who commanded naval air forces at Naval Air Station Halifax in Nova Scotia in the First World War, flew over the North Pole in a three-engine Fokker F.VIIA named *Josephine Ford*. Byrd and Navy Chief Aviation Pilot Floyd Bennett flew from Spitsbergen in a flight lasting fifteen hours and fifty-seven minutes, including thirteen minutes spent circling at their furthest north. When he returned to the United States from the

Lieutenants John Macready and Oakley Kelly with their Fokker T-2.

The *Josephine Ford*, the aircraft that Richard Byrd flew to the North Pole on his 1926 polar expedition.

Arctic, Byrd became a national hero. Congress passed a special Act in December 1926, promoting him to the rank of commander and awarding both him and Floyd Bennett the Medal of Honor. His flight has been marred by controversy since 1926, especially as in 1958 the Norwegian-American aviator and explorer Bernt Balchen cast doubts on Byrd's claim on the basis of his knowledge of the aeroplane's speed. Balchen claimed that Bennett had confessed to him months after the flight that he and Byrd had not reached the pole. The validity of the flight was never proved one way or the other, but in November 1929 he flew over the South Pole with Balchen. Byrd died as one of the most highly-decorated officers in the history of the United States navy. He is, probably, the only individual to receive the Medal of Honor, Navy Cross, Distinguished Flying Cross and the Silver Life-Saving Medal. He also was one of a very few individuals to receive all three Antarctic expedition medals issued for expeditions prior to the Second World War.

Alan Cobham was another British aviator with a thirst for long-distance air travel. In 1921 he made a 5,000-mile air tour of Europe, visiting seventeen cities in three weeks, but it was four years later that he flew from London to Cape Town and back in his DH.50. On 30 June 1926, he set off on a flight from Britain to Australia. Some 60,000 people swarmed across the grassy fields of Essendon Airport, Melbourne when he landed his DH.50 seaplane which had been converted to a wheeled undercarriage earlier at Darwin. During the flight to Australia, Cobham's engineer, Arthur Elliot, was shot and killed after they left Baghdad on 5 July 1926. In 1929 Cobham mounted his first tour of Britain called the Municipal Aerodrome Campaign. This was an ambitious plan to encourage town councils to

Alan Cobham flew a de Havilland DH.50 from London to Cape Town and in 1926 he flew to Australia and back. This photograph depicts the aircraft used by Cobham in his London to Cape Town flight.

build local airports in the hope of drumming up business for his activities as an aviation consultant. The tour visited 110 venues between May and October 1929 using a ten-passenger DH.61 Giant Moth named *Youth of Britain*.

The military aircraft of the period kept largely to the biplane format which offered performance at a time when airspeed was not considered high enough to

The DH.61 Giant Moth used in Cobham's tour of Britain.

A Bristol Bulldog.

exploit the cleaner lines of the monoplane. English fighter aircraft included the Fairey Flycatcher, Firefly and Fox, the Gloster Grebe and Gamecock and the Bristol Bulldog, culminating with the Hawker Fury in 1929 which had a top speed of over 200 mph. The Bulldog was, of course, the aircraft that a young Pilot Officer Douglas Bader made famous by crashing and losing both his legs into the bargain. Exceptions to the rule were made by the Bristol monoplane, an experimental mid-

A Hawker Fury II with streamlined wheel fairings.

wing machine with a retractable undercarriage, the French Dewoitine D.27 and the Junkers K.47, which eventually gave birth to the Junkers 87 Stuka.

In Britain two machines were produced: the Beardmore-Rohrbach RO VI Inflexible, a three-engine all-metal cantilever monoplane which suffered from being underpowered, and the Fairey Long-Range Monoplane. It was in the Fairey Long-Distance Monoplane that Squadron Leader Oswald Gayford and Flight Lieutenant Gilbert Nicholetts (navigator) broke the world distance record in a flight from Cranwell in Lincolnshire to Walvis Bay in South-West Africa during February 1933.

In 1919 the New York hotel-owner Raymond Orteig offered the $25,000 Orteig Prize to the first aviators to make a non-stop transatlantic flight between New York and Paris in the next five years. No one won the prize, so he renewed the offer in 1924. The major competitors for the Orteig Prize were a veritable 'Who's Who' of early aviation. They included René Fonck, a celebrated fighter pilot who had been the Allies' ace of aces during the First World War, and Richard Byrd, the American who won the Medal of Honor for flying over the North Pole. In a flight that probably transformed the entire picture of American air transport and private flying, Charles Lindbergh took up the challenge with his Ryan monoplane in May 1927 with a record-breaking flight of thirty-three hours, thirty-nine minutes. His

Charles Lindbergh standing next to the *Spirit of St Louis*.

A postcard of the Levasseur PL.8 called *L'Oiseau Blanc*. The insets are of Charles Nungesser (left) and François Coli.

flight in the *Spirit of St Louis*, which had a similar impact to Blériot's crossing of the English Channel in 1909, finally created the financial and technical climate in America for the large-scale development of aviation. Twelve days before Lindbergh's flight of 20-21 May was the scene of an attempt by Charles Nungesser, a First World War flying ace with more than forty aerial victories, and François Coli, another wartime veteran, in a Levasseur PL.8 called *L'Oiseau Blanc*. Taking off from Paris at 5.17 am on 8 May, the biplane was escorted to the French coast by four military aircraft led by French Air Force Captain Venson. A sighting was also made by the commanding officer of the British submarine HMS *H50*, who recorded in the submarine's log that the aircraft was seen 20 miles south-west of the Needles. After passing over Ireland, no further verified reports were made. The disappearance of *L'Oiseau Blanc* is considered one of the great mysteries in the history of aviation. Many rumours circulated about the fate of the aircraft and crew, with mainstream opinion at the time being that it was probably lost in a squall over the Atlantic. Inspired by Charles Lindbergh's successful transatlantic flight, James Dole, the Hawaiian pineapple magnate, offered a prize of $25,000 in May 1927 for the first fixed-wing aircraft to fly the 2,405 miles from California to Honolulu in Hawaii. The race, if one can call it that, was referred to as the Dole Derby and attracted eighteen official and unofficial entrants. Fifteen of those drew for starting positions and of those fifteen, two were disqualified, two withdrew and three aircraft crashed before the race, resulting in three deaths. Eight aircraft eventually participated in

The winning aircraft in the Dole Derby was a Travel Air 5000 called *Woolaroc* flown by Hollywood stunt pilot Arthur 'Art' Goebel and William Davis.

the start of the race on 16 August, with only two successfully arriving in Hawaii. Of the six unsuccessful aircraft, two crashed on take-off, two were forced to return for repairs and two went missing during the race. One of the aircraft that was repaired took off again to search for the missing aircraft several days later and also vanished over the sea. In all, before, during and after the race ten lives were lost and six aircraft were lost or damaged beyond repair. Despite the deaths, the race opened up the air travel business to Hawaii, but by December 1932 Dole was removed from management of the company.

On 29 June 1927, thirty-three days after Charles Lindbergh's record-setting transatlantic flight, Bert Acosta flew from Long Island to France with Commander Richard Byrd, Lieutenant George Noville and Bernt Balchen aboard the *America*, a three-engine Fokker C-2 monoplane. When *America* reached Paris after forty hours, a heavy fog prevented it from landing. Acosta and his crew flew around Paris for about six hours waiting for the fog to lift and, when it failed to do so, they flew west and landed in the sea off Ver-sur-Mer in Normandy on 1 July

Jim Mollison (centre) made the first east-to-west solo transatlantic flight from Ireland to Canada in a de Havilland Puss Moth.

1927. Byrd, Acosta, Balchen and Noville ended up entering France in a rubber life raft as *America* sank after landing on the water. It took until April 1928 for the more difficult east-to-west crossing of the Atlantic to take place by the Bremen, a German fixed-wing Junkers W 33 which was flown by Hermann Köhl, and four years later in 1932, Scotsman Jim Mollison made the first east-to-west solo transatlantic flight, flying from Ireland to Canada in a de Havilland Puss Moth. While gaining a reputation as a playboy, Mollison was a skilled pilot who, like many others, took to record-breaking as a means of making his name. In July 1931 Mollison set a record time of eight days, nineteen hours for a flight from Australia to England and in March 1932, a record for flying from England to South Africa in four days, seventeen hours.

Amelia Earhart was theoretically the first female aviator to fly across the Atlantic Ocean, although she only accompanied the pilot Wilmer Stultz and Louis Gordon on her first flight. Taking off on 17 June 1928 from Newfoundland in a Fokker F.VIIb, they landed in south Wales twenty hours later. However, on the morning of 20 May 1932 she completed the journey again from Newfoundland to Northern Ireland in her Lockheed Vega VB in a solo flight lasting nearly fifteen hours.

Amelia Earhart.

Lockheed Model 10E Electra similar to the type flown by Amelia Earhart.

On 11 January 1935, Earhart became the first aviator to fly solo from Hawaii to California. Although this transoceanic flight had been attempted by many others, notably by the unfortunate participants in the 1927 Dole Air Race that had reversed the route, her trailblazing flight had been mainly routine with no mechanical breakdowns. Sadly, during an attempt to make a circumnavigation of the world in 1937 in a modified Lockheed Model 10-E Electra, Earhart and her navigator Fred Noonan disappeared over the central Pacific Ocean near Howland Island.

Jim Mollison had flown commercially for Charles Kingsford Smith's short-lived Australian National Airways and during one of his flights he met the equally famous aviator Amy Johnson, who he married in July 1932. In February 1933 Mollison flew from England to Brazil in three days, thirteen hours, using Africa as a stop-over continent, a record time and the first solo crossing, and Johnson achieved worldwide recognition when, in 1930, she became the first woman to fly solo from England to Australia in a second-hand DH.60 Gypsy Moth named *Jason*. In May 1936, Johnson made her last record-breaking flight, regaining her Britain to South Africa record in a Percival Gull VI. She was awarded the Gold Medal of the Royal Aero Club in the same year. She was killed in 1941 when the RAF Airspeed Oxford she was ferrying crashed into the Thames.

In May 1928 Kingsford Smith, Charles Ulm, Harry Lyon and James Warner flew to Australia from California in the *Southern Cross*, a Fokker F.VIIb three-

Amy Johnson standing in front of her Gypsy Moth before she undertook a nineteen-day solo flight to Australia.

engine monoplane. Flying via Hawaii and Fiji, they covered a distance of about 7,250 miles.

In 1920 the Italian Arturo Ferrarin, along with First World War flying ace Guido Masiero, flew an Ansaldo SVA biplane from Rome to Tokyo in multiple stages and in July 1928 he and fellow aviator Carlo Del Prete flew a single-engine land aircraft Savoia-Marchetti S.64 flying boat from Rome to Brazil. The flight set the world distance record for a non-stop flight, but during the festivities in Rio de Janeiro, Ferrarin and Del Prete crashed during a demonstration flight in an S.62 in August and Del Prete died from his injuries five days later. On

Arturo Ferrarin.

his return to Italy from Brazil in 1928 Ferrarin was awarded the Gold Medal of Aeronautic Valour. He participated twice in the Schneider Trophy, flying a Macchi M.39 in 1926 and a Macchi M.52 in 1927. On both occasions he had to withdraw before the end of the race due to engine trouble (his bad luck is charted in Chapter 6). Ferrarin's death came, like so many of his contemporaries, in July 1941 when a new experimental plane he was testing crashed at Guidonia Montecelio, near Rome. Guido Masiero was killed on 24 November 1942 when he collided with another C.202 flown by Francesco Agello in heavy fog over Milan. Sadly, Agello was also killed.

Probably the most famous airliner in aviation history was the Douglas DC-3, also known as the Dakota, which was introduced in December 1935. While the Boeing 247 was the prototype for the Bristol Blenheim and inspired both the German Junkers 86 and the Heinkel He III, the DC-3 was faster, had wing flaps for better control and carried twenty-one passengers. Over the next six decades some 10,000 DC-3s were produced and today some are still flying.

The development of flying boats and seaplanes between the wars followed the same pattern in most countries with an aviation industry, the exception being Germany which was still technically bound by the terms of the Treaty of Versailles, although manufacture in Switzerland and Italy rather negated the impact of Versailles. Commercial transport still used adapted war surplus aircraft and only really survived thanks to government subsidies granted for carrying mail. Passengers were initially unable to pay the actual cost of travelling and were seen as additional baggage to the freight.

The DC.3 could carry up to twenty-one passengers.

The American aviator Lieutenant Commander Reid made his crossing of the Atlantic a couple of weeks before Alcock and Brown's historic flight. Flying a Curtiss NC-4 seaplane in May 1919, Reid was the first man to make the crossing of the Atlantic in stages, from the United States to Newfoundland, then to the Azores and Portugal where it flew on to Britain. The whole journey took twenty-three days but was not eligible for the *Daily Mail* prize since it took more than seventy-two consecutive hours.

General Billy Mitchell was the man behind the round-the-world flight by four American Douglas World Cruiser biplanes, which could be fitted with either floats or wheels. The original four expedition aircraft, named *Boston*, *Chicago*, *New Orleans* and *Seattle*, departed from Sand Point, Washington in March 1924. The lead aircraft, *Seattle*, crashed in Alaska on 30 April, while the other three aircraft, with *Chicago* assuming the lead, continued west across Asia and Europe. The *Boston* was forced down and damaged beyond repair in the Atlantic off the Faroe Islands, leaving the remaining two aircraft to continue across the Atlantic to North America and Newfoundland. The surviving aircraft returned to Seattle on 28 September 1924. They had flown 19,942 miles in approximately 371 hours at an average speed of 70 mph. Billy Mitchell was an outspoken pilot who first came to the notice of the powers that be in September 1918 when he planned and led nearly 1,500 British,

Colonel Billy Mitchell in the cockpit of a SPAD XIII during the Battle of Saint-Mihiel.

French and Italian aircraft in the air phase of the Battle of Saint-Mihiel, one of the first coordinated air-to-ground offensives in history. Rather like Italo Balbo of Italy, Mitchell was out to gain prestige for American aviation but sadly, his rather splendid intentions were marred by his constant disagreements with all and sundry, including the American President Calvin Coolidge. He resigned in February 1926 and spent the next decade writing and preaching air power to all who would listen. He died of a variety of ailments, including a bad heart and an extreme case of influenza in New York on 19 February 1936 aged 56.

While the Curtiss NC-4 and the Douglas World Cruisers were largely experimental aircraft, the Dornier Do-J Wal ('Whale') was manufactured in large numbers for both military and civilian purposes. Derived from a wartime aircraft, the Wal's characteristic feature was the lateral fin that was placed underneath the wings to prevent the aircraft rolling and the need for floats and its back-to-back twin engines. The Wal first flew from Spain to the Canary Islands in January 1924 with the Spanish pilot Ramón Franco at the controls. In 1926 Franco again took off and on this occasion flew to Buenos Aires in seven stages to complete the first east-to-west crossing of the south Atlantic. Ten years after the first flight Wolfgang von Gronau was responsible for flying a much-improved and modified aircraft from the Baltic to New York via Iceland, Greenland, Labrador and Montreal, a feat he repeated during the next year. Not content with that, he completed a flight around the world in November 1932.

The Savoia-Marchetti S.55 that first flew in August 1925 was intended to be a bomber and torpedo aircraft but was hijacked to some extent by General Italo Balbo as the ideal propaganda machine in which to promote the Italian *Regia Aeronautica Italiana* (Italian Air Force). In February 1927 Mussolini suggested that

The Dornier Do J *Wal* was a twin-engine German flying boat designed by Dornier *Flugzeugwerke*.

Francesco de Pinedo make a flight to the Western hemisphere to inspire pride in people of Italian ancestry who had emigrated to the Americas. This idea developed into the Four Continents flight of 1927, which was intended to demonstrate the ability of a flying boat to fly from Italy to Africa and across the Atlantic Ocean to Brazil. The flight would be followed by several stops in South America and the Caribbean, a tour of the United States and Canada, and a transatlantic flight back to Europe ultimately ending in Rome. It was an ambitious project, but one that would certainly inspire Italians worldwide if it could be accomplished.

Pinedo, his co-pilot Carlo Del Prete and mechanic Vitale Zacchetti took off in a Savoia-Marchetti S.55 flying boat called the *Santa Maria* from Sardinia in February 1927 and after numerous adventures, including a fire that destroyed the aircraft, they travelled by train to New York where they arrived on 25 April 1927 to meet a new S.55 shipped there by the Italian Fascist government so that they could continue their flight. The new aircraft, which was identical to the *Santa Maria*, arrived in New York by ship on 1 May 1927, and, after reassembly, was christened *Santa Maria II*. However, their adventures were not over and on 22 May, they departed Trepassey Bay in Newfoundland, planning to cross the Atlantic to the Azores, refuel and then fly on to Portugal, retracing the transatlantic route of the United States navy Curtiss NC-4 flying boat in 1919. However, running low on fuel due to unfavourable weather, Pinedo was forced to land the *Santa Maria II* on the ocean and be taken under tow by a Portuguese fishing boat and an Italian steamer for the final 200 miles to the Azores. After a week of repairs, the three Italian aviators were airborne again in the *Santa Maria II*, flying back to the point in the Atlantic where they had been taken under tow, and then finishing their transatlantic flight. After stops in Portugal and Spain, Pinedo, Del Prete and Zacchetti completed the Four Continents flight on 16 June 1927, landing the *Santa Maria II* in Ostia's harbour outside Rome. Pinedo's death came in September 1933 when he attempted to take off from New York in a specially-constructed Bellanca aircraft en route for Baghdad with 1,000 gallons of fuel on board. Losing control of the aircraft and unable to gain altitude, he veered off the runway and crashed. Thrown from the cockpit, he reached back into it to turn off the smouldering plane's engine but the fuel vapours ignited and he died in the resulting fire. Pinedo's body was burned beyond recognition in the fire.

However, it was the ambitious flights made by Balbo that placed the twin-hulled, twin-engine S.55 firmly on the map. On 6 November 1926, though he had only a little experience in aviation, Italo Balbo was appointed Secretary of State for Air and went through a crash course of flying instruction before embarking on building the *Regia Aeronautica Italiana*. Balbo relied heavily on Francesco del Pinedo's advice when planning and executing the mass formation flights, which were intended to

The Savoia-Marchetti S.55. All the passengers or cargo were placed in the twin hulls, but the pilot and crew flew the aircraft from a cockpit in the thicker section of the wing, between the two hulls. The aircraft was used extensively for Italian propaganda, namely the Four Continents flight of 1927.

improve the operational skills of *Regia Aeronautica Italiana* aircrews and ground crews, showcase the Italian aviation industry to potential foreign buyers of Italian-made aircraft, and enhance the prestige of Benito Mussolini's Italian Fascist government, a cause for which Balbo became famous. Such was his organizational power that he was made General of the Air Force in August 1928 and, ten months later, Minister of the Air Force. It was Balbo who restructured and took charge of the Italian Schneider Trophy team, but perhaps his greatest exploit was the flight between Rome and Chicago with a squadron of twenty-four S.55Xs in July 1933. The flight began with a crossing of the Alps to reach Chicago on 15 July, returning to Italy via the Azores on 12 August. Balbo had shown with his series of flights that seaplanes were technically and practically capable of making long flights across oceans and the progress made by Italian seaplanes was in keeping with the original aims of Jacques Schneider, although it is doubtful if Mussolini and Balbo considered Schneider's long-term views for more than a moment. In 1933, perhaps to relieve tensions surrounding him in Italy, Balbo was given the government of Italian Libya, where he resided for the remainder of his life. Early in the Second World War he was accidentally killed by friendly fire when his plane was shot down over Tobruk.

In Britain, flying boats became essential in the connection of the major cities of the British Empire. However, despite three consecutive Schneider Trophy wins, Supermarine was never able to capitalize on these successes and in the flying boat routes operated by Imperial Airways. Shorts appeared to be the dominant manufacturer, a tribute perhaps to the all-metal structure pioneered by Oswald Short. Shorts' direct competitor was Supermarine, which built the Southampton in 1925, a twin-engine biplane flying boat with the tractor engines mounted

Alan Cobham selected the Short Singapore for his Round Africa flight in 1927. This photograph shows a Short Singapore III in the process of taking off.

between the wings. Critically the Southampton Mark I had both its hull and wings manufactured from wood and it wasn't until a year later that Mitchell used an all-metal structure for the Southampton Mark II.

One of the first signs of Shorts' dominance was the selection of a Short Singapore flying boat by Alan Cobham for his 1927 Round Africa flight. Being of all-metal

The Short 23 flying boat photographed in 1938. The aircraft was known as the C-Class to Imperial passengers and was developed and manufactured in parallel with the Short Sunderland.

construction, the Singapore was in no danger of the joints in the flying surfaces being weakened by exposure to water and humidity. In 1928 a Short S.8 Calcutta replaced the Supermarine Swan on the Channel Islands service and appeared again on the route to the Indies, which was opened up on 30 March 1929. In fact the Short Calcutta was the first of the Imperial Airways flying boats to be built and flew from London to Karachi on the first through air service between Britain and India. Later in the same year this route was extended to Jodhpur and Delhi. The Calcutta was replaced by three Short S.17 Kents in May 1931. Imperial Airways also used the Short S.23 and S.26 on their Brindisi to Alexandria route.

It is fitting that this short appreciation of the development of flight between the wars should end with Charles Lindbergh, who became an enthusiastic seaplane pilot for Pan American which was founded in 1927 as a scheduled air mail and passenger service operating between Florida and Cuba. In February 1929 Lindbergh made the first mail flight between Miami and the Panama Canal in a Sikorski S.38, a twin-boom amphibian that could carry 1 ton of mail or nine passengers. In April 1930 he flew in the S.38 again, this time from Miami to Buenos Aires and back. Incidentally, the S.38 began life as a landplane in which Lindbergh established a new speed record between California and New York.

Chapter Five

The Schneider Trophy 1919–1922

'If 1919 perpetuated the farce which had so often characterized the early hydro-aeroplane meetings, 1920 and 1921 degenerated into fiasco. In all three contests an Italian pilot finished alone.'

Ralph Barker, writing in *The Schneider Trophy Races.*

Having looked briefly at the development of flight between the wars, let us return to the Schneider Trophy contests to see what influence they had on aviation. Although the First World War had provided British aviation with a considerable boost, marine aviation had not fared so well, a situation reflected in the 1919 Schneider Trophy. Contrary to public acclaim there was no Gordon Bennett Race in 1919 and Howard Pixton's win in 1914 meant that the Royal Aero Club was required to stage the event in Britain. Anticipating a large British entry, the Royal Aero Club announced that the contest would be run at Bournemouth in early September. The starting and finishing line was off Bournemouth pier with turning-points at the Old Harry Rocks and Hengistbury Head to complete a course of 20 nautical miles. As far as spectators were concerned, the ten laps of the course gave them ample opportunity to see the turning-points from the cliffs.

The Sopwith Schneider.

In early September the British entrants, now reduced to a disappointing four biplane machines, presented themselves for the eliminating trials. High on the list was the Sopwith Schneider, a much-improved aircraft from 1914 and the only British entry that had benefited from a protracted period of development in that it had operated as a seaplane during the war. However, critics felt it was still a landplane at heart, despite its new powerplant, the 450hp Cosmos Jupiter air-cooled radial engine, which was neatly cowled in a very streamlined appearance crowned with a large spinner, but in Harry Hawker's experienced hands perhaps it looked like a winner.

The Fairey IIIA beached at Bournemouth.

Even though it was specially prepared by the company to compete in the contest, the Fairey IIIA was in effect another converted landplane. With its wingspan reduced to a mere 28ft and powered by a 450hp Napier Lion engine, it was flown by Lieutenant Colonel Vincent Nicholl and was expected to test the Sopwith for speed. Nicholl was later to become the man behind the Fairey Swordfish, the aircraft that in May 1941 led the famous attack on the *Bismarck* and crippled her steering gear, leaving her to the mercy of the Royal Navy surface ships.

The Supermarine Sea Lion I.

The Supermarine Sea Lion was a flying boat, a dedicated sea-going design powered by the Napier Lion 450hp engine. Flown by Canadian Squadron Leader Basil Hobbs, who was decorated with the DSO for shooting down Zeppelin L.22 in his Curtiss H.12, the machine appeared to favour rugged seaworthiness over speed. The Sea Lion was a somewhat conventional single-bay aircraft with a mahogany hull, wingtip floats and a pusher propeller arrangement. Designed by Frederick 'Jim' Hargreaves, Reginald Mitchell's predecessor, many thought that it stood little chance against the lighter aircraft with tractor propellers. The last of the British contenders was the Avro 539A, which was essentially a small biplane built around the 240hp Siddeley-Deasy Puma six-cylinder in-line liquid-cooled engine and flown by Australian Captain Harold Hamersley. A former ace with 60 Squadron, he shot down ten German aircraft and destroyed another three driven down out of control. The Avro was remarkably similar to the Sopwith Schneider in that it was a small single-bay biplane mounted on twin floats with a wingspan of some 25ft, although it did differ from the Sopwith machine with its large frontal radiator through which the propeller shaft extended. The trials on 3 September effectively ruled out the Avro's chances when a float was damaged on take-off and although the aircraft was given another chance to show off its worth five days later, it failed to dislodge the Sea Lion from its third place. Hamersley was placed in reserve.

The French entrants focused on the SPAD-Herbemont XVII flown by Joseph Sadi-Lecointe and powered by a 340hp Hispano-Suiza liquid-cooled engine. Sadi-Lecointe was another individual who had served as a pilot during the First World War. Active in the French Resistance during the Second World War, he died in July 1944 as a result of being tortured by the Gestapo in Fresnes Prison.

The Avro 539A with a wheeled undercarriage, although it was originally designed and built as a seaplane.

The two Nieuports were variants of the standard Nieuport 29 and had already damaged themselves en route to Bournemouth. The first, flown by the Corsican Lieutenant Jean Casale, was wrecked after hitting a mooring buoy at Cowes, and the second, flown by Henri Mallard, went down in the Channel, leaving him to cling to the wreckage for twenty-four hours until he was rescued. Jean Casale came to prominence in 1915 when he moved to *Escadrille* 8 as a pilot flying Maurice Farmans. By December 1916 he was officially an ace, shooting down sixteen enemy aircraft and one balloon, and was one of the few French aces that survived the entire course of the war. Casale was killed in a flying accident on 23 June 1923 while flying a four-

Sadi-Lecointe standing in front of his Nieuport-Delage 29V after winning the Gordon Bennett Trophy at Orléans on 28 September 1920.

engine Blériot 115. However, reputation alone was not enough to put his Nieuport back in the running and, despite setting new height and speed records earlier in the summer, Casale's aircraft was declared a non-starter, a fate which was also to cloud the SPAD-Herbemont as it too was declared a non-starter.

The Italian entry centred on the specially-built S.13 flying boat flown by the blond Sergeant Guido Janello of the Italian Air Force. Powered by a 250hp

Jean Casale was killed flying the prototype of the four-engined Blériot BL 115 in June 1923.

The Savoia S.13. The pilot in the cockpit is probably Sergeant Guido Janello.

Isotta Fraschini engine, it was a sleek reconnaissance aircraft that had its wing area reduced to increase its speed. The actual challenge came from the Savoia Company, whose founder and president was Lorenzo Santoni. Not unknown to the British establishment, he had been the first to institute the British Aeroplane Syndicate in 1911. In April 1912 Santoni, along with Frenchman Maurice Prévost, put up a record by being the first pair to fly from Paris to England in the day to deliver a 70hp Gnome-powered Deperdussin to the Admiralty. The machine was the first foreign machine ordered by the Admiralty and Santoni had determined that it should be delivered by air. In November 1920 he established the *Chantiers Aéro-Maritimes de la Seine* (CAMS) and in 1921 began production of Savoia types S-9, S-13 and S-16 under licence. However, as far as Bournemouth and the 1919 Schneider Trophy was concerned, Santoni was convinced that the aircraft would do well in the forthcoming contest.

The meeting was beset by incompetent organization and the vagaries of the British weather, all of which added up to a farcical meeting which was eventually declared null and void. Blanketed in a late summer fog, the race was repeatedly postponed with the machines only making a few circuits of the course before retiring. The only exception was Guido Janello in the S.13, who completed the required ten laps but unfortunately failed to observe that he had been rounding a spare marker boat that had been anchored not far from the official one. His subsequent disqualification produced a storm of protest and the Royal Aero Club did the only decent thing and recommended that Italy be awarded the trophy.

However, the *Féderation Aéronautique International* (FAI), throwing up its hands in dismay at the appalling organization, ruled the flight was invalid and rather pointedly awarded the next venue to Italy.

* * *

Accordingly, the fourth contest was held east of Venice in September 1920 and again it was something of a failure, despite the Italians wishing to do things according to the rules. Wishing to at least improve on the Bournemouth fiasco, the *Aero Club d'Italia* was convinced that the more agreeable climate of the Adriatic coast would be more conducive to competition, but in this they were to be sorely disappointed. Most countries were still feeling the effects of the First World War, American participation was a non-starter, Germany was constrained by the Treaty of Versailles and Britain and France withdrew citing the prohibitive costs of entering. Nevertheless Italy pressed on with their plans, but even they were embroiled in domestic issues and their original entry of four machines was eventually whittled down to one. On the day of the race the S.12 flying boat, powered by a 550hp Ansaldo engine and flown by Lieutenant Luigi Bologna, finally completed the course with an average speed of 107.2 mph. The fourth contest was thus a fly-over for Italy.

* * *

The fifth contest was also held in Venice on the same course as the previous year during August 1921. It was equally unsatisfactory in that there was no British entry and the single French entry, a Nieuport-Delage powered by a 300hp Hispano-Suiza engine and flown by Sadi-Lecointe, crashed before the event. Sadi-Lecointe hit the sea rather heavily during landing in the navigability trials and buckled the float structure, thereby putting his machine out of the contest. There remained only the three Italian entries selected for the race: two Macchi 7s and one Macchi M.19. Both Arturo Zanetti in one of the M.7s and Pietro Corniglio in the M.19 were forced to retire, leaving Giovanni de Brigante in the second of the M.7s to complete the ten laps in an average speed of 117.8 mph. It had been an Italian fly-over again.

* * *

For three years Italy had monopolized the trophy and it looked very much as if 1922 would see the splendid *Coppa Schneider* reside permanently in Italy; indeed,

A Macchi M.7 taking off from Lake Roxen in Sweden. The Swedish flag can just be made out on the tail assembly.

without financial help from the government the economic situation in Britain was about to see a British challenge fall at the first hurdle. However, all was not lost as the Supermarine Company was about to mount a serious challenge to save the trophy from the clutches of the Italians, albeit with a fair amount of co-operation from the British aviation industry. Mitchell's promotion to chief designer in 1919 brought him into the world of designing flying boats to his own specification and the Sea Lion II was his answer to the rather precarious situation in which the British now found themselves. One of the unexpected events that influenced Mitchell's career as a designer was the third and final win by Sadi-Lecointe in the Gordon Bennett Cup of 1920. As with the Schneider Trophy, the third consecutive win gave permanent possession to the winning country and, more importantly, left the Schneider Trophy as the only remaining international speed contest. Naturally this suited Supermarine extremely well as it focused their attention on seaplanes.

The Sea Lion II was to be flown by Henry (sometimes spelt Henri) Biard and one can only imagine his expression when he was told by Hubert Scott-Paine that his aircraft was to be a remodelled version of the 1919 Schneider Trophy entry. To Biard it must have looked more or less like an outdated version of a flying boat that had little chance of winning the forthcoming contest. As chief test pilot for the Supermarine Company he flew the first commercial cross-Channel air services from Woolston to Le Havre with Supermarine's offshoot, British Marine Air Navigation, and was one of the first, if not *the* first, to pilot a commercial flight to

The Sea Lion II at the Supermarine works at Calshot.

the Channel Islands. He also had a wartime pedigree, being commissioned in 1917 in the RNAS, and flew a number of anti-submarine patrols in a Wright converted seaplane.

Biard's initial observations had reckoned without the genius of Reginald Mitchell, who had been set a target speed of 160 mph by Scott-Paine. Substituting the Hispano-Suiza engine for the 450hp Napier-Lion, Mitchell set to work

Reginald Mitchell, the designer behind the Sea Lion II.

streamlining the fuselage and reducing the wing area. In response to the additional power of the engine, he modified the vertical surfaces above the tailplane, giving the leading edge a pronounced forward curvature. Although the finished machine was based on the Sea King, it was given the name Sea Lion II, drawing attention to the name of the engine that Harry Vane of the Napier Company had donated. The *Aero Club d'Italia* announced that the event would take place in August among the beautiful surroundings of the Bay of Naples, the rules would be similar to that of the previous year and the course would consist of thirteen laps of the 15.5- mile circuit. Dominated by Mount Vesuvius to the east, it was a triangular course starting and finishing just west of the harbour with a sharp turn at Torre del Greco. The contest promised to provide the excitement missing from the two Venice contests.

France had gone all out in support of a new company established by Lorenzo Santoni, formerly the managing director of the Savoia Company and a man we had briefly met during the 1919 event. Founding CAMS in 1920, he had gone into business with the Savoia designer Raffaele Conflenti, producing aircraft with a distinct similarity to the Savoia machines! Nevertheless, the two French ones were based on the CAMS 31 flying boat, a pusher machine with a reduced wing span and a 300hp Hispano-Suiza engine and renamed the CAMS 36. Sadly they were withdrawn from the race, leaving the British to fight it out with the Italians.

The challenge to Henry Biard's Sea Lion II came from a new design from the Italian *Società Idrovolanti Alta Italia* (SIAI). The S.51, powered by a single 300hp Hispano-Suiza engine and flown by Alessandro Passeleva, looked every inch a contender for the trophy with its streamlined hull and sesquiplane configuration, a compromise between the monoplane and biplane constructions, retaining enough

The S.51 was flown by Alessandro Passeleva and placed second in the race.

The rather ungainly M.17 flown by Arturo Zanetti was little match for the Sea Lion II.

of the lower wing to keep the biplane structure and ensuring that a large part of the upper wing could function aerodynamically as a monoplane.

The two Macchi aircraft were an M.17 flown by Arturo Zanetti and powered by a 260hp Isotta-Fraschini engine, and an M.7bis flown by Piero Corgnolino which, with a similar power unit, completed the Italian line-up. As expected, the Italian team were more than a little anxious to gain some knowledge of just how fast the Sea Lion II was. They reasoned that unless a minor miracle was to occur

The Macchi M.7 in Swedish colours.

the machine posed little threat but, perhaps remembering the 1914 event, it was as well to be prepared. Biard, using his gamesmanship, purposefully never allowed the engine to display its full potential during the air tests. To what extent the Italians were deceived is unclear, but what was clear was that the S.51 looked the faster machine and the experts were forecasting an Italian victory. There was a near disaster during the navigability trials when the S.51 took in water and capsized. According to the rules, the S.51 should have been disqualified but, in a gesture of pure sportsmanship, Scott-Paine raised no objection to it being salvaged and it duly lined up the next day. However, but for fate taking a hand, he may well have lived to regret this decision.

A delay on race day until late afternoon, largely because of the heat, jangled all the competitors' nerves until Sea Lion II finally lifted off at six minutes past the hour. Biard opened the throttle and found himself at the first turn by the Cote di Posillipo almost before he had settled down. He had complete confidence in the Napier-Lion engine as he knew the carburettors had been especially tuned for the high temperatures encountered at that time of day. Cutting the corners and banking hard, his first lap time proved to be the fastest lap of the race by a margin of seven seconds. By the time he had finished his first lap, Corgnolino and Zanetti had taken off and joined the circuit and realized that the Macchis were little match for the Sea Lion II; all now depended on Passeleva in the S.51.

However, Passeleva had discovered problems with his machine that would eventually prove to be his undoing. Despite closing on Sea Lion II and very nearly taking pole position, the ducking the aircraft had undergone the previous day had affected the glue bonding the laminations of the wooden propeller. This in turn caused the aircraft to vibrate, forcing Passeleva to throttle back. Now he could only concentrate on the task in hand and hope that Biard would make an error of judgement or his engine would break down under the strain. For the spectators it provided all the excitement missing from the previous two years. Unaware of Passeleva's problems, they only

Henry Biard qualified as a pilot in a Howard-Wright biplane in 1912. Undoubtedly he would have celebrated his win at Supermarine and may have even joined Mitchell in Southampton.

knew there was very little between Sea Lion II and the S.51. To those keeping the lap times they watched as the S.51 slowly gained on the British machine. Would the Italian be able to steal a victory from Sea Lion II?

In the event it was a close-run thing, but Biard was the winner with an average speed of 145.7 mph, Passeleva was just over two minutes behind him in the S.51 with an average speed of 143.20 mph and Zanetti came in third with the M.17. Trailing in last place was Corgnolino in the M.7, which only achieved an average speed of 123 mph. There is little doubt that the damage sustained by the S.51 in the navigability trials cost the Italians the permanent possession of the trophy in 1922, but it should be remembered that if it hadn't been for Scott-Paine's sportsmanlike attitude, the S.51 would not have flown at all if the rules had been adhered to. Reginald Mitchell was finally able to taste the sweet contentment of success with his design of the Sea Lion and on return to Southampton enjoy the spectacle of the city dignitaries turning out in full ceremonial dress to celebrate Supermarine's success.

The ill luck that had haunted the Savoia Company time and time again had once more proved to be their undoing. Their S.51 could easily have captured the trophy; in fact, during a flight in December 1922 the aircraft recorded a speed of 174.08 mph and captured the world speed record for seaplanes. There was little doubt that the Italians were still favourites to gain permanent possession of the trophy in the 1923 contest at Cowes.

Chapter Six

The Americans Join the Fray: 1923–1926

'The 1923 Schneider contest was the final proof that America or, more accurately, the Curtiss Company, was now building the fastest, most reliable high-speed aircraft in the world.'

Don Vorderman in *The Great Air Races.*

The British government may well have been persuaded to follow the example of the French and provide some financial backing to Supermarine and the other aviation firms had they known just how the American challenge was to be conducted. In America, rather than leaving the entry to the various firms designing and building the aircraft, they turned over responsibility to the United States navy; they in turn approached the contest with military precision, bringing with them an imposing array of equipment and support crews on the USS *Pittsburgh*, which anchored off Cowes. The best the British government could do was to offer to purchase the winning machine, providing of course it was British, for £3,000, less the engine. They could not be moved to increase their offer, especially in the light of the prevailing economic conditions and were apparently blinkered to the view that the benefits to British aviation would be enormous if a national aircraft were to win the trophy in 1923.

By mid-July there was still no sign of a British entry and it came almost as a relief to the Royal Aero Club to learn of a Supermarine entry, the construction of which obviously spurred on two further contenders from Blackburn and Hawker, prompting the Royal Aero Club to announce the final details for the event in late September. The start and finish lines of the course were at Cowes opposite the Victoria pier and the circuit was to consist of five laps of a 37-mile course giving a total distance of 186 miles with turning-points, identified by marker boats, at Southsea and Selsey Bill. The rules stipulated that the starting line had to be crossed on the surface and the finishing line while in flight, repairs at sea were permitted during the race without assistance and refuelling was only allowed if surplus fuel was carried in the aircraft. During the speed trial the team for each country would take off together separated by an interval of fifteen minutes.

Not all the three aircraft of the British entry got to the start line: of those, the only machine that ranked as a seaplane was the Sopwith/Hawker Rainbow, a small

The Americans Join the Fray: 1923–1926   101

The Sopwith/Hawker Rainbow was a rebuild of the Sopwith Schneider by the Hawker Company with the float undercarriage removed and with a wheeled undercarriage substituted. It was initially powered by an ABC Dragonfly engine (later a 500hp Bristol Jupiter was installed). In that form it was retained by the Hawker Company for air racing until it crashed on Burgh Heath golf course on 1 September 1923. The photograph depicts the Sopwith Schneider before the rebuild.

biplane that had finished second with an average speed of 164.5 mph in the Aerial Derby six weeks earlier. Powered by a 500hp Bristol Jupiter radial engine, it was flown by Flight Lieutenant Walter Longton, another flying ace of the First World War credited with eleven confirmed aerial victories. Longton unfortunately crashed on Burgh Hill golf course during a test flight and although he was miraculously unharmed, the aircraft was withdrawn.

The Blackburn Pellet was powered by a 450hp Napier Lion engine and flown by Lieutenant Commander Reginald Kenworthy. Almost withdrawn because of

The Blackburn Pellet.

The Sea Lion III. Henry Biard is standing in the cockpit.

the disastrous ducking the aircraft had in the Humber estuary, it was salvaged and packed off to Cowes with only days to spare. Sadly the Pellet failed at the very beginning of the navigability tests when it swerved to avoid a rowing boat, causing the machine to porpoise and sink after becoming momentarily airborne. Fortunately Kenworthy managed to escape after a nightmare sixty-one seconds beneath the water. The third of the British entrants was the Sea Lion, a machine that had won the trophy the year before at Naples. With the realization that they would have little chance of beating the Italians unless the performance of the Sea Lion II could be substantially improved, Reginald Mitchell had set to work early in the year to squeeze the last mile per hour out of the aging flying boat. Flown once again by Henry Biard, it was powered this time by a 550hp Napier Lion engine, the same engine that had been fitted to the Gloster Bamel that had won the Aerial Derby that year with an average speed of 193.4 mph. The engine manufacturer Napier succeeded in coaxing another 100hp out of the engine and Mitchell reduced the drag to a minimum, modifying the nose of the hull and improving the design of the wings. In this new form the Supermarine entry was designated Sea Lion III.

The Italian challenge dramatically fizzled out when the hoped-for offer of assistance from their government came to nothing. Clouded by economic and political difficulties, there was little hope for a team that had hoped to win the trophy and place it permanently in Italy. For the 1923 contest there only remained the Savoia S.51, which had broken the world seaplane record in December 1922, but Alessandro Marchetti, Savoia's managing director, realizing that increased power must come from an improved engine, was certain that his company would

A *Chantiers Aero-Maritimes de la Seine* (CAMS) 36bis single-engine flying boat.

be throwing money away if no suitable powerplant was available. Ultimately the economics were prohibitive and Savoia withdrew, forcing Italy out of the contest.

Meanwhile the French, aware that they had only won the trophy on one occasion, were determined to see it back at the *Aéro Club de France.* Consequently every possible support was given to the French team of flying boats consisting of two CAMS, two Lathams and one Blanchard. The CAMS 36 that was flown by Lieutenant Pelletier d'Oisy was basically the same aircraft that had been entered for the 1922 contest except for the new 360hp Hispano-Suiza engine; with this addition the aircraft was redesignated the CAMS 36bis. Perhaps the showpiece of Santoni's factory was the CAMS 38 flown by Maurice Hurel and sporting a modified tail unit and a streamlined hull. Powered by a 360hp Hispano-Suiza engine with a two-bladed pusher propeller and a reduced wing area, it retained the overall dimensions of the CAMS 36. The two Latham flying boats, powered by two 400hp Lorraine-Dietrich twelve-cylinder engines, we are told, relied on

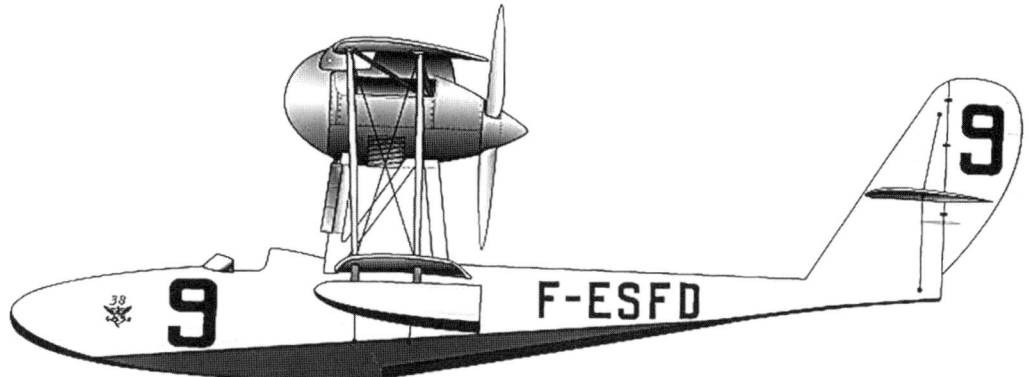

The CAMS 38 was flown by Maurice Hurel.

The Blanchard C1 was withdrawn from the race.

seaworthiness rather than speed, a 'characteristic' that had already reduced their number to one. The surviving Latham, flown by Alphonse Duhamel, was also a non-starter in that it lined up on the start line ready to take to the air when a sudden backfire sheared a magneto drive. It was largely the same story with the Blanchard C1 flown by Paul Teste, a French naval officer who was killed in June 1925 at Villacoublay on an Amiot 120 during a training flight for an attempt at crossing the Atlantic. Teste joined the *Aviation Navale* in 1917 and was shot down on 26 May and taken prisoner in Karlsruhe. After a first unsuccessful escape attempt, he managed a 'home run' in January 1918. The Blanchard C1, powered by a 400hp Bristol Jupiter engine, was a parasol-wing monoplane of gigantic proportions. When tested it could only achieve about 150 mph, a speed that was nowhere near fast enough and was consequently withdrawn. With its wings placed above the fuselage, the aircraft was certainly distinctive compared to the popular biplanes of the time. It was now up to the CAMS flying boats to take on the American Curtiss racers.

Among the pressures operating behind the scenes in America was the appointment of General Billy Mitchell who, as assistant chief of the Army Air Service was attempting to create a separate Air Force and increase American interest in air racing. An intense rivalry grew up between him and Rear Admiral William Moffett, which acted rather like a tonic to the aircraft industry, resulting in Congress granting sufficient funds for the construction of a number of racing aircraft. In 1921, in order to facilitate the development of military fighters, the United States navy entered into a contract with the Curtiss Aeroplane and Motor Company for the production of pursuit biplanes. Buoyed by this initial success, Billy Mitchell succeeded in extracting further funds from Congress for development, which in turn spurred the Curtiss Company into producing twelve new aircraft for the 1922 Pulitzer contest. Several months before the Pulitzer race,

The Americans Join the Fray: 1923–1926   105

A Curtiss CR-3. The pilot in the cockpit may well be Lieutenant Rutledge Irvine.

Curtiss submitted new designs to the navy but the navy, not wanting to order new aircraft, turned them down. Never slow to seize an opportunity, Billy Mitchell entered into a contract with Curtiss to develop two machines designated the R-6. Finishing first and second in the 1922 Pulitzer race, the two R-6s demonstrated to the world that they were ready for any challenge to their supremacy. The American government was out to gain prestige for American aviation and to stimulate research

Navy Wright NW-2.

and development from which it was hoped a marked improvement would result in standard service aircraft.

Therefore, great interest was shown in the American entry which arrived in late August in order to familiarize themselves with the course and water conditions. The chosen aircraft were the two original racers built for the navy in 1921 and the second prototype of a new navy racer built by the Wright Aeroplane Company. All the aircraft were seaplanes and comprised two Curtiss CR-3 racing machines, the navy Wright NW-2 seaplane and a Naval Aircraft Factory TR-3A that was used for practice. Potentially the most formidable was the NW-2, a biplane with a fuselage and tail unit made of a welded steel-tube construction and sheet aluminium skin from the nose to the pilot's cockpit. Powered by a 700hp Wright T-3 twelve-cylinder in-line liquid-cooled engine, it was probably the fastest single-engine seaplane in the world. Interestingly, the propellers of all three aircraft had blades made of forged duralumin, a development that eventually became the norm as speeds increased.

The NW-2, flown by Adolphus 'Jake' Gorton, proved to be trouble-free in practice, leaving Gorton sure that his aircraft would win the trophy, but calamity was just around the corner. A change in the fixing of the propeller during the aircraft's last test flight resulted in a near-disastrous crash causing the NW-2 to hit the water and catapult Gorton clear at nearly 200 mph. Quite unbelievably he was uninjured,

Jake Gorton was involved in a crash that effectively ruled the NW-2 out of the race.

Like Gorton, Lieutenant Frank Wead was ruled out of the race after the TR-3A refused to start.

but the aircraft was out of the race, placing the responsibility of winning onto the two CR-3s. These machines, known as Curtiss navy racers, were formidable-looking aircraft, one of which had been flown by Bertrand 'Bert' Acosta and won the 1921 Pulitzer Trophy with an average speed of 176.7 mph. To help offset the drag imposed by the large floats needed for the 1923 contest, both aircraft were subject to increased vertical tail surfaces and their engines were upgraded to the Curtiss 465hp CD-12 direct drive. In this new form they were redesignated CR-3. What was perhaps not immediately obvious were the surface radiators built into the wing sections to transfer maximum heat from the engine: these radiators created little or no drag, which was the direct opposite of conventional radiators mounted in the slipstream. Some observers even maintained that the more conventional radiator created sufficient drag to absorb as much as 10 per cent of engine power. The Americans also suffered another casualty. Lieutenant Frank 'Spig' Wead, who had been designated team reserve, was now ready to fly in the TR-3A and make up the team of three when the starting gear sheared; the aircraft refused to start by swinging the propeller and Wead was ruled out of the race. However, it was not the last we would hear of Wead. In 1924 he and Lieutenant John Price, using a Curtiss CS-2 with a Wright T-3 Tornado engine, set new seaplane records for distance and duration. Frank Wead, after serving with the Pacific fleet during the Second World War, retired in 1945 to become a screenwriter.

Race day dawned in the warmth of the summer sun and the two Americans, Lieutenants Rutledge Irvine and David Rittenhouse, crossed the start line and roared away. Spectators at Selsey Bill were quite sure that Rittenhouse looked to be slightly in the lead with Irvine close behind him. Biard was next to take off and was staggered to see the two Americans flash past him overhead; if lap times were anything to go by then their speed was well beyond Sea Lion III's capability. Joining the circuit, he tried every trick in the book to coax the machine along and catch up at the crucial turning-points. Misfortune also clouded the French team, particularly as only the CAMS 38 flown by Maurice Hurel was able to take off after crossing the start line. The CAMS 36bis collided with a moored yacht and damaged its hull, enough to put it out of the race. Meanwhile, Hurel took nearly twenty minutes over the first lap and on the second he was forced to set down with engine trouble off Selsey Bill. Now only three aircraft were contesting the trophy.

Even to the most uninformed spectator it was clear that the result was a foregone conclusion; that is, barring accidents! Indeed, far from faltering, the American aircraft improved their lap times on each occasion they passed the finish line. Biard, however, not to be outdone, did the same and his average speed over the fifth and final lap was in excess of 160 mph, a speed that would have won the 1922 contest but was not enough to win on this occasion. Rittenhouse's final lap

Lieutenant David Rittenhouse seen standing on the float of his Curtiss CR-3 after winning the 1923 Schneider Trophy contest.

averaged 181.1 mph, and with a race average of 177.38 mph he was the clear winner over Irvine who averaged 173.46 mph. Biard's 152.17 mph, albeit faster than the previous year, left him trailing in third place.

* * *

It must have been very clear to Mitchell and Supermarine that an aircraft with an impressive engine cooled by flush-fitted wing radiators was going to be very hard to beat. The European flying boat with an engine mounted above the fuselage had, for racing purposes, seen its day. On top of this, it had probably dawned on Mitchell that the Curtiss CR-3 limited the number of drag-reducing items to sixteen struts and twenty wires, whereas the Sea Lion had thirty-three struts and forty-two wires. The American design also allowed for an engine to be neatly cowled and merged into the fuselage.

The fame of the Curtiss racer was now established: in winning the Schneider Trophy it had also flown faster than it did when winning the 1921 Pulitzer Trophy with a wheeled undercarriage. It had been predicted that if the Americans won the Schneider Trophy at Cowes it would never again be brought back to Britain, and it now looked very much as if that prediction would be proved to be true and there was little to stop them winning the trophy outright. Certainly the Italians were relieved that the British had been beaten and denied a second victory, preventing

Two R2C-1s were entered in the 1923 Pulitzer Trophy race with Lieutenants Alford Williams and Harold Brow taking first and second places. This photograph shows Lieutenant Alford Williams leaning against the fuselage of an RC2-1.

them from an outright triumph in 1924. At least the only nation that could win the trophy outright in 1924 was Italy, but as if to demonstrate to the aviation world that the Curtiss aircraft were the fastest, eight days after their victory at Cowes, the United States navy took the first two places in the 1923 Pulitzer Trophy at St Louis in two new Curtiss biplanes that had been specially built for the race. The winning Pulitzer pilot was Lieutenant Alford Williams, who achieved an average speed of 243.67 mph. Even though the prospect of a European challenger beating the Curtiss racers in the twelve months before the 1924 Schneider Trophy was looked upon as almost impossible, Britain, France and Italy all entered the next contest regardless. Scheduled to be hosted by the Flying Club of Baltimore at Bay Shore Park, Chesapeake Bay in October, the upbeat attitude of the American team masked the problems that were being experienced in Italy. The problem of failing to find a suitable engine was dogged by a very limited design experience and political turmoil, exacerbated by the rise of Mussolini in 1922. The French, never really a contender for the trophy, withdrew along with the Italians, leaving the Gloster II as the only European aircraft challenging the Americans. The Gloster II was very similar in appearance to the Gloster Bamel and it was thought that, given the right circumstances, the machine would prove to be a real challenge to the American Curtiss racers.

Five weeks from the race, Hubert Broad took the Gloster II for a test flight, and although he considered the aircraft to be slightly tail-heavy, it managed a creditable

Hubert Broad.

200 mph. Satisfied, he touched down and just as everyone was congratulating themselves on the initial test, the aircraft began to porpoise, resulting in the float struts giving way under the strain and the whole machine sinking, leaving only the tail unit above water. Initially thought to have drowned, Broad managed to swim clear and was rescued while clinging to the tail. Britain was out of the contest. With no time left to prepare another aircraft, it looked very much as if the American entry would run away with the trophy with a fly-over.

The American team of two Curtiss racers was to be supplemented by a navy Wright machine designated the F2W-2, which was powered by a Wright T-3 engine and flown by a very cynical Jake Gorton, who confided later that he never really trusted the aircraft and was furious with himself for going up in it at all. His initial assumptions were quite correct and during a test flight in October the aircraft crashed on landing, hitting the water upside-down. Gorton miraculously managed to escape and was picked up by a passing tugboat. The crash brought to an end to the Wright racing aircraft and Gorton never again flew in a Schneider Trophy contest.

It was with considerable relief that the anticipated fly-over by the American team never took place, thanks to a generous degree of sportsmanship by the American National Aeronautic Association. They wrote to the Royal Aero Club, and presumably the national aviation bodies of France and Italy, to say that because of the withdrawal of Italy and the accident suffered by the British they felt that it would be in everyone's interest to cancel the race for 1924. By allowing a period of twelve months in which overseas competitors could prepare, the Americans had made one of the most charitable gestures in the history of the Schneider Trophy and, to the cynics among us, at the same time demonstrated their dominance over the rest of the aviation world with the Curtiss racers. In fact, after the race was cancelled, a seaplane meeting was scheduled at Bay Shore Park in October where the two CR-3s raced over the course originally established for the 1924 contest. Lieutenant George Cuddihy, flying the previous year's winning Curtiss CR-3, established a number of records in the process, including one run over 3 kilometres at 188.1 mph.

* * *

After Hubert Broad's ducking in the Gloster II it was realized that Britain had a great deal to learn about high-speed aeroplanes. Mitchell's Sea Lion III was no match for the American machines and even the Blackburn Pellet flying boat had been noticeably more streamlined than the entry by Supermarine. Once more it was the lottery of external conditions that made Mitchell's design of the Supermarine S.4 a reality; the unexpected events provided by the postponement of 1924 allowed Mitchell to produce a dedicated and competitive racing machine. As was seen with the design of the Spitfire, his first response was not as dramatic as first thought, but it was the cantilevered monoplane that evolved from this early thinking that enabled him to break away from the flying boat in favour of the new Supermarine design. Mitchell was clearly intent on the next step which would render the Curtiss racers obsolete and his answer was a machine more than a decade ahead of its time. To this day the S.4 is considered by many to be one of the most beautiful aircraft ever built and it is reported that when Henry Biard, the pilot who was to fly it, first saw the S.4 he was awestruck:

> In 1925, to eyes accustomed to the galumphing flying boats of the period, with their clutters of struts and wires, the stark white monoplane mounted on its two sleek floats seemed almost sinister. It had wings like great knife blades jutting out from the fuselage – not a bracing wire anywhere – and the tiny cavity for the pilot was set so far back it seemed almost at the tail.

Powered by a 700hp Napier Lion engine, the fuselage was a monocoque structure with the front of the machine constructed from steel tube to which the wings and undercarriage struts were attached. Landing flaps were fitted in an attempt to overcome the faster landing and take-off speeds that were characteristic of a higher wing loading.

The S.4 is considered by many to be the most beautiful aircraft to have ever been built.

However, there were drawbacks to Mitchell's design: visibility, according to Biard, was poor and on a take-off or landing the pilot couldn't see at all. On more than one occasion Biard was lucky not to collide on the water and while in the air, in fact, during the first test flight in August 1925 disaster nearly overtook the S.4 owing to the pilot's cockpit position. The aircraft's centre of gravity dictated that the pilot's position should be *behind* the trailing edge of the wings; this in its turn

A side view of the S.4 showing the poor visibility experienced by the pilot.

created a blind spot and Biard almost crashed into the SS *Majestic* while taking off. Biard also thought he detected a slight flutter in the wings: 'Immediately I got into the cockpit I knew there was trouble coming in that machine. It didn't feel right. Besides, the visibility from the cockpit was perfectly dreadful. The wings were in the way.'

Biard's confidence in the aircraft did not increase with time and although it was clearly fast, he hid his fears concerning the flutter in the unbraced single wing, putting it down to imagination. During the next few weeks he flew the S.4 on several more occasions and on one occasion at Calshot achieved 226.75 mph, a speed that set a new seaplane record, beating the existing record established by George Cuddihy in a Curtiss racer. Nevertheless, with the two Macchi M.33s not presenting too much of a problem, was this the year that Supermarine would overcome the Americans to emerge victorious?

Hubert Broad, in the meantime, was managing problems of his own. On loan from de Havilland, he only managed to take the Gloster III into the air on 22 August and discovered to his horror that the sliding and skidding course the aircraft seemed determined to make was the fault of directional stability. Time did not allow the redesign that was really necessary, but the derisory tail surface was temporarily rectified, bringing about a noticeable improvement. A second Gloster III had been built as a reserve machine and was to be flown by Squadron

The Gloster III.

Leader Bert Hinkler who served initially with the RNAS as a gunner/observer in Belgium and France and was awarded the Distinguished Service Medal before becoming a pilot. In 1918 Hinkler concluded his war with 28 Squadron in Italy. After the war he undertook a number of long-distance flights including the non-stop flight in an Avro Baby to Bundaberg, Queensland, a distance of 850 miles. During the 1920s he flew non-stop from England to Latvia, for which he was awarded the Oswald Watt Gold Medal.

It was a measure of the hurried nature of the programme that Hinkler only flew the Gloster III on one occasion before the machine was packed up en route for America. The site chosen by the Americans for the 1925 race was Bay Shore Park just south-east of Baltimore, but owing to bad weather it was postponed until late October. The course was a triangular one over Chesapeake Bay starting and finishing at Bay Shore consisting of seven laps of 27 miles over a total distance of 189 miles. There were turning-points at the lighthouse south-east of Gibson Island and a pylon just west of Huntington Point.

The Italian team, which left Genoa aboard the SS *Conte Verde*, was sponsored not by the government but by *Aeronautica Macchi*. The only help they received from the government was the loan of two Curtiss D-12 engines which General Alessandro Guidoni had purchased the previous year. The engines had been so extensively tested by Fiat that, despite being incorporated into the two Macchi M.33s, they were never to produce their full power. In designing the Italian aircraft

The Macchi M.33.

Mario Castoldi had turned, like Mitchell had done with the S.4, to the cantilever monoplane, except he did not abandon the flying boat structure altogether. The engine was mounted on struts above the hull and wing, giving the whole machine the look of a flying boat. The machine was not well liked by the pilots; in particular they complained about frequent vibration of the cantilever wing, a similar problem that Biard had with the S.4. The two M.33s were flown by Giovanni de Briganti, the winner of the 1921 race, and Ricardo Morselli.

The Americans purchased four newly-designed Curtiss machines powered by Curtiss V-1400 engines that were expected to deliver 600hp. One of the new airframes was tested to destruction, leaving three aircraft and a spare engine, and of these, two were entered by the United States navy and one by the army. The new Curtiss seaplanes were designated R3C-2 and were flown by Lieutenants George Cuddihy and Ralph Ofstie of the United States navy and army Lieutenant James Doolittle. Cuddihy was killed in November 1929 while testing a Bristol Type 105 Bulldog II fighter at Anacostia Naval Air Station. Doolittle is probably most famous for leading the secret attack of sixteen B-25 medium bombers against some of the Japanese main islands from the aircraft carrier USS *Hornet* in April 1942. He was subsequently awarded the Medal of Honor. He concluded his war in command of the Eighth Army Air Force in Britain.

However, disaster was soon to strike the British team. First the inclement weather dumped a heavy tent pole onto the tail of the S.4 causing considerable damage and

The Curtiss R3C-2 racing aircraft. The pilot is thought to be Lieutenant Ralph Ofstie.

second, a bout of influenza effectively reduced Biard to a sweating, shivering lump. He had already suffered the misfortune of breaking a bone in his wrist during the voyage to America, prompting the British team leader to suggest that Bert Hinkler flew the S.4. Realizing that his experience of flying the S.4 was invaluable, Biard declined, and on the morning of the preliminary trials climbed into the tiny cockpit and gave the signal to stand clear. Ofstie, de Briganti and Broad had already taken off. Biard was fast off the water and circling over the pier head he was seen to make a steeply banked turn. Was he in trouble? Seconds later he hit the water:

> A roar that seemed to shatter the world as my lovely monoplane hits the water and exploded on the impact like a bomb, an impression of green water, solid and hard as ice through which I am flung like a spear through snow, a horrible agonizing pain in all my limbs and through my body, and I regain consciousness 30ft down under icy motionless sea water, legs in the air, head and shoulders pinned down under awful glue like mud, scrabbling and dragging with broken fingernails and paralysed, frozen fingers at the safety strap that still binds me helplessly to a splintered piece of what was formerly the cockpit.

To the spectators it looked as if Biard had gone down with the aircraft and drowned, but as Broad approached the crash site in his Gloster III, a bedraggled shape was seen to emerge on the surface. Biard had survived and freed himself from the wreckage. A rescue launch was soon on the scene and among the occupants was Mitchell, a man who must have been extremely worried for Biard and at the same time devastated at the loss of the S.4. Subsequently Biard was found to have two broken ribs and considerable muscle damage and in the investigation at Supermarine following the crash it was suggested that the accident may have been caused by wing flutter in the cantilever wing. Another theory was put forward that the crash was caused by the 'indisposition of the pilot as a result of the physical effects of flying at very high speed.' Perhaps Biard should not have been allowed to have his own way; it was later argued that he was in no fit state to handle a racing aircraft and that Hinkler should have flown the aircraft, but we shall never know the truth of the matter. Whether torsional weakness or a stall accounted for the crash, wing flutter cannot be ruled out, but what was certain was that with the increase in flying speed this characteristic needed to be fully understood and remedied.

The loss of Mitchell's S.4 meant that Bert Hinkler and the reserve Gloster III had to be prepared to take part in the race. Therein lay another problem: the facilities at Baltimore were just not capable of getting the second Gloster ready in time and despite a unanimous appeal by all the competing pilots and the inclement weather, Hinkler was eventually eliminated after the float structure on the Gloster collapsed. Britain's only hope now lay with Hubert Broad.

The speed competition began at 2.30 pm on 26 October with Doolittle the first to take off, followed by Broad, Cuddihy and Ofstie, and de Briganti in the M.33 bringing up the rear. It soon became clear that the small, modest Doolittle was putting up the fastest lap times, which was hardly unexpected as he had the reputation of being a brilliant pilot who was among the first to combine theory and practice in research flying. His course in aeronautical engineering had taken him into the realms of the effect of high centrifugal and gravity (G-force) on pilots. His experience now showed as his tight turns around the marker pylons were in direct contrast to the slower and wider turns made by the navy pilots. Broad, who had 100hp more than the R3C-2s, attempted to emulate the tight turning of Doolittle, but the directional stability on the Gloster III left much to be desired. Broad later said that his aircraft was behaving like the back wheels of a car skidding on an icy road. The differing cornering techniques must have been an impressive sight for the spectators, but as was so often the case with the Schneider Trophy entrants, time was short in which to sort out problems with the aircraft.

Then disaster hit again, and this time it was the American team that was on the receiving end. Ralph Ofstie was forced down with engine failure on lap six and George Cuddihy's engine caught fire on lap seven, leaving only three aircraft challenging for the trophy. To the relief of the American spectators, Doolittle's

Jimmy Doolittle standing on the starboard float of his RC3-2 Curtiss racer.

engine maintained its speed and as he crossed the finish line he recorded an average speed of 232.573 mph. He was followed by Hubert Broad, who managed 199.17 mph, and de Briganti in third place with 168.44 mph.

Mitchell had already come to the conclusion that the flying boat was outdated and his design of a monoplane without bracing wires of any sort had been a surprise, albeit an unsuccessful one. On the other hand, a special place should be reserved for the S.4 in the history of the Schneider Trophy and indeed in the design career of Reginald Mitchell. It was left to Ermanno Bazzocchi of *Aeronautica Macchi* to write that 'the real revolution of 1925 was the appearance of the Supermarine S.4: its very clean design set the pattern for all the subsequent Schneider racers.'

* * *

The call for some sort of high-speed flight of pilots who had ample time to fly the machines that were now being entered for the Schneider Trophy was being echoed in the corridors of the Air Ministry and Royal Aero Club. The victorious American service pilots and their back-up crews were being cited as the way forward; despite the skill of civilian pilots and their willingness to take risks, their usual flying experience was in much slower machines. More to the point, the isolated flying experience of the Schneider machines was exacerbated by the limited amount of practice time that was usually available. Schneider events were typically characterized by late decisions, shortened flying time at race sites and either poor weather or mechanical problems. Furthermore, it was not known until the FAI had met at the beginning of each season whether there would be any changes to the race regulations.

Certainly the American team of service pilots with their Curtiss racers had changed the nature of the Schneider Trophy forever and made the previous preparations for an annual contest not only uneconomic but impractical. Only government sponsorship along the lines enjoyed by the Americans could produce the machines capable of making a successful challenge. Much to their credit, neither Britain nor Italy considered abandoning the trophy, but did agree that a postponement until 1927 would be to everyone's advantage. However, what had not been taken into account was the attitude of the Americans: since their postponement of the race in 1924 there was a stubborn determination to run the race in 1926 and get the whole scenario over and done with. It would seem that the Americans had no intention of relaxing their grip on the trophy and intended to win for the third and last time. Thus, in December 1925 the Americans advised all potential contenders that the race would be held at the Naval Air Station at Hampton Roads, Virginia in the week beginning October 1926. In late January 1926 the Royal Aero

Air Vice Marshal Sir William Sefton Brancker was a British pioneer in civil and military aviation and senior officer of the RFC and later the Royal Air Force. He was killed in an airship crash in 1930.

Throughout the 1920s Mussolini remade Italy in his image and, seizing an opportunity to promote his Fascist regime, backed the Italian entry in the 1926 Schneider Trophy contest.

Club informed the American National Aeronautic Association that Britain would not be competing in 1926, adding a request for a formal postponement until 1927. It fell on deaf ears. The reaction in Britain was profound disbelief, forcing Air Vice Marshal Sir Sefton Brancker, the chairman of the Royal Aero Club's Racing Committee, to personally sail to America in late February to make a final appeal. It came as something of a rude shock that Brancker's appeal was rejected. What was even more of a surprise was the news that an entry had been received by the Americans from the *Aero Club d'Italia*, overturning the entire British strategy for getting a postponement. As there was now a challenger, there would be a race and the Americans looked likely to win; the trophy and the status it commanded would be theirs for all time.

What the British did not know was that the Italian entry was in reality backed by the government and, furthermore, Benito Mussolini had ordered the race to be won. Understandably Macchi and Fiat were the two firms chosen by the dictator and while the M.33 had signalled a movement towards the design of the S.4 monoplane, they now went further, abandoning the flying boat hull and producing instead a monoplane floatplane. Macchi's designer, Mario Castoldi, went one step further still and produced the M.39, a distinctive and elegant aircraft with a low wing and cruciform tail unit looking much like a cross. The 38-year-old Castoldi

Mario Castoldi designed the M.39 monoplane, an elegant breakaway from the standard flying boat hull.

had moved to *Macchi Aeronautica* at Varese in 1922 where his intuitive and original designs soon singled him out as a perceptive aircraft designer. During the Second World War Castoldi took charge of the design of a series of military fighters that formed the mainstay of the Italian fighter force, specifically the C.200, C.202 and Macchi C.205.

In the intervening time Fiat, and in particular Tranquillo Zerbi, having learned from their in-depth study of the Curtiss D-12 engine, produced a twelve-cylinder V engine that was capable of producing in excess of 880hp. In mid-June the Italian entries were confirmed and consequently there was no question of a postponement; the first machine was ready for testing at Lake Varese by late August with Romeo Sartori, Macchi's chief test pilot, at the controls. The remaining two aircraft were delivered to the Schiranna Air Station on Lake Varese where the team of Italian pilots chosen to fly in the race assembled. Under the wing of the 26-year-old Vittorio Centurione, the Italian team consisted of Major Mario de Bernardi, Captain Arturo Ferrarin and Lieutenant Adriano Bacula. Ferrarin was the chief test pilot of Fiat and had already flown in the record-breaking Rome-Tokyo flight in 1920 and would go on to participate in the non-stop flight from Italy to Brazil in 1928 with fellow aviator Carlo Del Prete. Ferrarin died on 18 July 1941 at the age of

46 when a new experimental plane he was testing crashed at Guidonia Montecelio, north-east of Rome. De Bernardi had been a First World War fighter pilot credited with four confirmed Austro-Hungarian kills. Like Ferrarin, he died in April 1959 when flying an aircraft, but on this occasion he managed to land the aircraft, only to die minutes later.

All appeared to be well with the team until on 21 September when Vittorio Centurione stalled on his first flight in one of the new M.39s and crashed into the lake; the aircraft was lost and Centurione was tragically killed. His replacement was Major Aldo Guglielmetti, but it is unclear whether he was party to the request by the Italians for a short postponement of two weeks while the team recovered from the death of Centurione. The Americans granted the request, which was again a generous one, especially as the British queried the decision on the grounds that a postponement beyond a day was illegal under the rules.

The three American machines were the identical Curtiss R3C-2 aircraft that had defended the trophy in 1925, except in two cases they were modified to accept two new engines. The third machine, flown by Doolittle in 1925, retained its 600hp Curtiss V-1400 engine. The new engines were a 700hp Packard, redesignated R3C-3, and

The three American machines were identical to the Curtiss R3C-2 racers that had defended the trophy in 1925, although in two cases they were modified to accept new engines.

Lieutenant Cyrus Bettis.

Lieutenant Hersey Conant.

Lieutenant Harmon Norton pictured with Gunnery Sergeant Esterbrook in 1919.

a 700hp Curtiss V-1550 redesignated R3C-4. Flown by three of the United States navy's best racing pilots, they looked almost certain to win. The American team was announced in September and was headed by Lieutenant George Cuddihy, the pilot who had been forced to drop out on the last lap in 1924, Lieutenant Frank Hersey Conant and Lieutenant William 'Red' Tomlinson. Lieutenant Harmon Norton of the United States Marine Corps was named as reserve.

Three weeks earlier, Lieutenant Cyrus Bettis, the winner of the 1925 Pulitzer Trophy and Doolittle's understudy during the 1925 Schneider Trophy contest, died of spinal meningitis after crashing his aircraft. With serious injuries including a broken leg and multiple skull fractures, he crawled 2.5 miles to a road where he was found by road workers. He was admitted to Bellefonte Hospital and then airlifted to Walter Reed Hospital in Washington. He died on 1 September 1926. This was compounded by the death of Harmon Norton on 13 September, whose aircraft crashed into 6ft of water on the Potomac River after completing a high-speed practice run. To a great extent this paralleled the death of Centurione on Lake Varese a week later and went some way to understanding the agreement of the National Aeronautic Association to a temporary postponement. All the same, we should not forget that the postponement certainly suited the Americans as well!

The loss of Bettis and Harmon was a severe blow to the American pilots, but it was to get worse. Lieutenant Hersey Conant, who had just established a new unofficial American record at Anacostia with a speed of 251.5 mph, was killed on 30 October while flying across the shallow waters of Winter Harbour while on his way to oversee the unpacking of his aircraft. Norton's replacement was Lieutenant Frank Schilt of the United States Marine Corps and that of Hersey Conant was the 30-year-old Lieutenant Carleton Champion, but the hard facts were plain to see: the Americans had lost half their team in the space of approximately six weeks, events that would surely impact on the outcome of the contest.

The Italian team, arriving from Genoa on the SS *Conte Rosso*, soon settled in at the Naval Air Base at Hampton Roads, a venue that was a huge improvement on the rather primitive facilities at Baltimore. Although the low-winged monoplane Macchi 39s drew admiring glances from the American spectators, there were numerous subtle hints concerning plagiarism, presumably drawing comparisons with the Supermarine S.4. With both teams suffering from mechanical difficulties, the actual test flying was again limited, all competitors appearing cautious about giving away too much concerning their speed. Consequently, the Italians were kept in the dark about the American potential and vice versa, although in reality there was little separating the two teams in terms of aircraft.

However, there were further setbacks. De Bernardi, the Italian captain, took off to complete his navigability trial on 11 November, but punctured his starboard float on landing when the motor launch sent out to tow him in collided with his aircraft. Later in the afternoon Arturo Ferrarin took off in the rapidly failing light, but failed to reappear by dark. Suffering an engine failure in the form of a broken connecting rod, he ended up drifting about on the surface of the water in imminent danger of being run down in the darkness by a passing ship. Spotted from the air by a service flying boat, the pilot landed on the water and taxied round the stricken Ferrarin for the next two hours. Finally, the rescue launch found him and he was towed back to the Naval Air Station. Working all through the night, the Italian mechanics managed to produce one serviceable engine and Ferrarin took off at 4.00 pm the next day to complete his navigability trial. If that was not enough, Tomlinson, flying the Packard-engined R3C-3, found the opposite torque of the Packard engine a little more than he had bargained for and bounced badly on take-off. After a test flight that appeared to suggest all was well, he misjudged his height on the approach, dropped heavily and damaged the starboard float, plunging the aircraft upside down in the water. Tomlinson fortunately struggled clear, but the R3C-3 was out of the race. After Ferrarin had completed his test, Tomlinson appeared again and took off in a standard Curtiss F6C-1 Hawk, demonstrating an awesome display of flying that had the spectators on the edge of their seats. The Hawk could not compare with the M.39s of the Italian team, but at least it was reliable and could be counted on to complete the course.

The triangular course was to be flown in an anti-clockwise direction beginning and finishing at Hampton Roads with turning-points in Chesapeake Bay and

Tomlinson was eventually placed fourth in the Curtiss F6C-1 Hawk. This photograph depicts an F6C-1 with a wheeled undercarriage.

Newport News. The spectators, of which there were said to be 30,000, could view the race from the comfort of Newport News, Hampton Roads as well as Old Point Comfort, situated between Newport and the turning-point in Chesapeake Bay.

The race was due to begin in the early afternoon and George Cuddihy was optimistic at the thought of the Americans winning. The first away was Bacula, followed by Tomlinson and Cuddihy with Ferrarin, de Bernardi and Schilt making up the last three to become airborne. The slowest, as expected, was Tomlinson in the Curtiss Hawk but even he was astounded by the speed of Bacula's M.39 which achieved 209.58 mph. If this was typical of the Italian aircraft, then the Americans had the race sewn up! However, any thoughts of victory were brought down to earth by Ferrarin's first lap, which was faster than Cuddihy, followed by de Bernardi who staggered them with a lap speed of 239.44 mph. Then, on lap four, Ferrarin's repaired engine failed, the cause being a fractured oil pipe, and the competition settled down to a contest between Cuddihy and de Bernardi with Schilt and Tomlinson bringing up the rear. Then disaster struck the American team: on the seventh lap Cuddihy dropped out of the race, just as he had done in the previous year. On this occasion his petrol supply had failed and despite the frantic pumping to lift the fuel supply

Lieutenant Frank Schilt of the United States Marine Corps with his Chance Vought O2U-1 Corsair.

Mario de Bernardi won the 1926 Schneider Trophy contest with a speed of 246.49 mph. This photograph shows de Bernardi in the cockpit of the M.33 after breaking the world speed record.

from the floats, his race was over. With the enforced retirement of Cuddihy, it was Frank Schilt who chased him home with an average speed of 231.36 mph. In the meantime de Bernardi, ensconced in his scarlet M.39, flew across the finishing line achieving an average speed of 246.49 mph followed by Schilt and Bacula. Tomlinson, flying the Curtiss Hawk, finished the race as predicted but could only average 136.95 mph. A telegram was immediately sent to Mussolini containing the words: 'Your orders to win at all costs have been carried out.' As if to cement their victory, de Bernardi confirmed their win four days later by setting a new world speed record of 258.87 mph.

Mario de Bernardi is carried aloft after winning the Schneider Trophy for Italy.

One must feel for George Cuddihy who had twice dropped out of the race on the final lap, but first is first and the Italians had finally produced an aircraft that ended the American domination of the trophy. By doing so they had ensured that the competition would continue, but the human cost had been high. Not only that, but both Macchi and Fiat had been completely disorganized by the production of aircraft for the Schneider Trophy contests and as a result work on other contracts had been put on hold. Yet the Italians did not forget their own: when the team eventually got back to Italy a wreath was dropped from a Macchi flying boat at the spot where Centurione had died on Lake Varese.

Chapter Seven

British Domination: 1927–1929

'The designing of such a machine involved considerable anxiety because everything had been sacrificed to speed. The floats were only just large enough to support the machine and the wings had been cut down to a size considered just sufficient to ensure safe landing. The engine had only five hours' duration; after that time it had to be removed and changed. In fact everything had been so cut down it was dangerous to fly.'

Reginald Mitchell speaking about the S.5
after winning the trophy in 1927.

In an attempt to solve the problem of whether the Schneider Trophy should be an annual or biennial event, an extraordinary meeting of the FAI was called in Paris in January 1927. Of the three countries involved, Britain, America and Italy – there seemed little chance of France participating – America favoured the biennial option, Italy, anxious to capitalize on de Bernardi's victory at Hampton Roads, wanted the contest to continue in 1927 and Britain, which had taken no part in the 1926 contest, was happy to try for the trophy in 1927. The two to one vote in favour of an annual meeting was enough to sway the FAI and it was announced that the next contest would take place on the Lido near Venice between September and November 1927. However, behind the American vote was a more sinister viewpoint. Having got what they wanted out of the competition and very much aware of the huge cost involved if they planned to win it again, they sought to downgrade the contest and reduce it to an international sporting competition, similar to yachting's America's Cup. America, it appeared, was more interested in a new venture called the National Air Tours which was designed to encourage an interest in air transport.

Since 1923 all the successful teams, notably the Americans, had been organized and manned by service personnel and if Britain was to win in 1927, they would have to do the same. Consequently, when the RAF's first High Speed Flight was formed on 1 October it was an admission that future success in the Schneider Trophy depended upon just that. Of course, this was another reason why Britain had voted in Paris for a continuation of the trophy to 1927; they simply did not wish all their preliminary effort wasted. The pilots assigned to the High Speed Flight

The National Air Tours were one of the crowning achievements of the Golden Age of Flight in America. Between 1925 and 1931 the National Air Tours for the Edsel Ford Reliability Trophy took place to promote aviation advancements in both aircraft technology and aviation infrastructure. This photograph is of the 1928 Ford National Reliability Air Tour taken on 22 July 1928 at Froid, Montana.

were mostly transferred from the Marine Aircraft Experimental Establishment at Felixstowe, others were volunteers who were selected and transferred in February. The final line-up, headed by Squadron Leader Leslie Slatter, consisted of Flight Lieutenants Samuel Kinkead, Sidney Webster, Oswald Worsley and Flying Officer Harry Schofield.

With regard to Mitchell, he must have been delighted with the formation of the High Speed Flight, especially since events had conspired to present him with the platform on which to create the S.5 and its successors the S.6 and S.6B. The propeller tip damage to Passeleva's aircraft in 1922, the American cancellation of the race in 1924, the demise of the Gordon Bennett land races and Mussolini's intervention in 1926, all contributed to the Schneider Trophy being the only international aviation contest and the consequent domination by Mitchell-designed aircraft.

The Italian team, clearly out to win the trophy again, was equipped with a development of the Macchi M.39. Mario Castoldi had excelled himself again with the design of the M.52, a monoplane design, smaller in length, with swept-back wings to accommodate the backward movement of the centre of gravity, which in turn had reduced area flying surfaces because of the power of the 1,000hp Fiat engine. This engine allowed a smaller frontage to be designed but the reliability of the Fiat engine was still questionable, largely because of the use of alloys and that

The Macchi M.52 was a monoplane design, smaller in length than the M.39 but with swept-back wings.

the higher compression ratio had perhaps not been fully researched. Nevertheless, the Italian team began training on Lake Varese in early summer to find that the stalling characteristics of the M.39s were rather delicate and, while de Bernardi and Ferrarin took this in their stride, Lieutenant Silvio Borra, a new pilot to the team, was killed in June during a practice flight.

British failure in 1925 prompted the Air Ministry to draw up specifications for three different racing machines and, following extensive appraisal involving

Gloster IV.

wind tunnel and tank testing, three designs were accepted: those of Gloster, Shorts and Supermarine. Obviously the Air Ministry was not following the Italian and American reliance on one type of machine. The Gloster design was surprisingly a biplane. Despite the Italian monoplane configuration that had won in 1926, Henry Folland still favoured the biplane, arguing that a well-streamlined biplane was well suited to the stresses and strains of aviation racing. So the Gloster IV was born. The new wing layout was more sesquiplane than biplane with the resultant reduction in wing area possible because of the geared Napier Lion engine producing 875hp. Streamlining was furthered by removing the top wing pylon mounting and replacing it with a more careful fairing into the top of the fuselage, as had been achieved with the Curtiss machines since 1925.

The Short entry was a low braced monoplane with a composite structure fuselage called the Short-Bristol Crusader. A fuselage frame consisting of high-tensile tubes provided the rigid attachment points for the wing spars and the four tubular struts that carried the floats, but the most unusual feature was the engine, a Bristol Mercury I nine-cylinder air-cooled radial type, a colossal departure from the water-cooled in-line engines that had powered the last three Schneider Trophy winners. Perhaps more worrying were the sudden engine cut-outs that Webster's flying log reported as a 'whip that nearly took it out of the machine'. One is driven to ask if

Bristol Crusader.

The S.5. This photograph was probably taken at Calshot before the 1927 contest.

the designers were aware that radial engine seaplanes stood little chance against the sleeker in-line engine machines, or if indeed Air Ministry interest was really in the capture of the trophy, particularly as the engine could only muster 650hp at best and individual helmets were fitted to the protruding nine cylinders as a compromise between cooling and streamlining.

At Supermarine, Mitchell continued to place his faith in the newer monoplane approach and strove for improvements to the design of the S.4. The cantilevered

The S.5 with Flight Lieutenant Oswald Worsley standing in the cockpit. Other members of the RAF High Speed Flight are seen at ground level looking on.

wing of the S.4 was replaced by a low wing which improved the pilot's view and afforded the best possible alternative that required the minimum of bracing. A wire bracing system between the floats, wing and fuselage was introduced along with a smaller fuselage and floats and a reshaping of the Napier Lion engine to give better streamlining. Apart from the increased horsepower of the Napier engine, the most telling improvement was a change from the Lamblin radiators of the S.4 to a system more akin to the Curtiss racers and the Macchi M.39. Together these improvements meant that a speed increase of 70 mph could be expected, bringing the top speed of the aircraft to somewhere in the region of 300 mph. In terms of composition the S.5 was of mixed metal and wood, with the all-metal fuselage being of a stressed skin structure while the flying surfaces were much like those of the S.4, of wooden construction and plywood-covered. Whatever Mitchell's personal thoughts were about stepping back from the cantilevered wing of the S.4, he had taken the more pragmatic approach to bracing and at the same time reduced the drag from the sturdy float struts we saw on the S.4. When the aircraft eventually arrived in Venice, the result of all the modifications was seen by the Italians as a direct copy of Castoldi's M.39, a criticism that failed to take into account the trend-setting S.4 of 1925 or the length of time Mitchell had spent on the design process. It is unclear whether Castoldi is on record as commenting on the S.5.

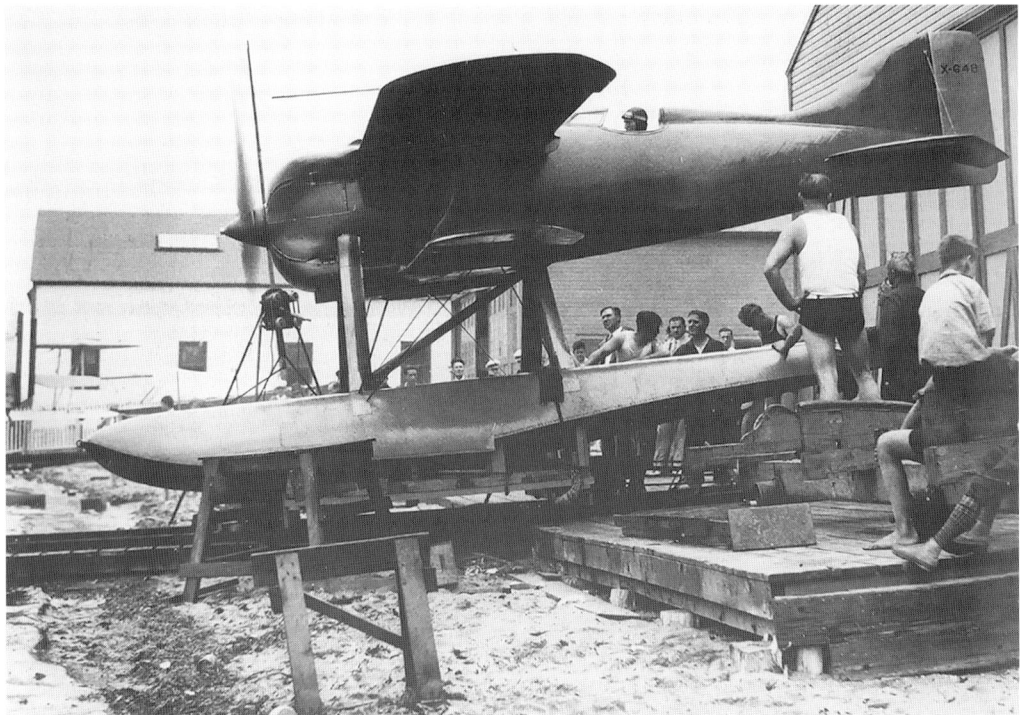

The Kirkham-Packard machine that was to have been flown by Al Williams before it was withdrawn.

After the withdrawal of government funding in America there was a private entry backed by a group of New York sportsmen with the sole intention of winning the Schneider Trophy. The pilot was to be Lieutenant Alford Williams, the winner of the 1923 Pulitzer Cup, who had obtained leave from the United States navy in order to compete in a specially-prepared Kirkham-Packard machine. Williams was not the only individual who thought that, given equal power, the Curtiss biplane should have beaten the Italian Macchi M.39 monoplane in 1926 and many thought the margin of victory was narrow enough to support this view. However, the Kirkham-Packard aircraft was experiencing delay after delay to the extent that the Americans requested a thirty-day postponement, rather like the Italians had done in 1924. Passing the request over to the British with the expectation that they would agree, it was rejected on the grounds that the cost involved of keeping the British team abroad for a month longer than expected was prohibitive, and anyway the rules did not allow for such a postponement. The rejection caused a political storm, but the Americans finally agreed that their request had been a little unreasonable, although the resentment continued for sometime afterwards.

The course in Venice covered 50 kilometres with the start and finish line in front of the Excelsior Palace Hotel, where the British pilots were being accommodated. Each of the seven laps had turning-points at Porto di Malamocco, Chioggia and Porto di Lido, near to the San Andrea Naval Station. Venice has been heralded as one of the finest Schneider courses, and from the viewpoint of spectators the

A section of the crowd along the waterfront at Venice for the 1927 Schneider Trophy Race.

beaches along the course offered magnificent vantage points, particularly along the mole and from the Excelsior Hotel.

With most of the British aircraft assembled in Venice, the British team commenced test-flying their aircraft. On 10 September Slatter took off in the S.5 on its first test flight in Italy. After a few turns over San Andrea he covered the full Schneider course, followed by Kinkead in the Gloster IV who made a low-level run along the Lido beach. The next day, after Webster had flown the S.5, Schofield prepared for his first test flight in the Crusader. Schofield was well used to the Crusader but he particularly disliked being clamped into position by the cockpit hood, likening it to being 'nailed inside a coffin'. As Schofield headed out into the lagoon there was a thump as a gust of wind hit the starboard wing, causing the wing to drop abruptly. Attempting to correct the tilt and finding that the wing had gone beyond the vertical, the Crusader hit the water, breaking in two. Schofield fortunately escaped with his life, but the post-mortem discovered that the crash had been caused by crossed aileron wires, the penalty perhaps of not giving one's full attention to the pre-flight checks! Out of the competition and nursing his injuries, he probably did not notice the arrival of the aircraft carrier HMS *Eagle* and its accompanying four destroyers that brought the remaining British aircraft; he was too intent on mastering the two crutches he was given as an aid to his mobility.

The navigation and watertightness tests were held on 23 September and all, as expected, completed them with the exception of Webster, who had to make a second attempt as he was judged to have crossed the start line incorrectly. Poor weather intervened and it was not until 26 September that the contest was programmed to begin after lunch. It was with some relief that Kinkead finally opened the throttles of the Gloster IV and took to the air, followed by Webster and de Bernardi. The new boy of the Italian team, Captain Federico Guazzetti, was next, then Worsley and finally Ferrarin. With the new rules demanding that each competitor now cross the start line while airborne, Ferrarin's race was almost immediately curtailed by his engine protesting loudly and then emitting smoke and flames. Seeking the security of a landing, he made for the sheltered waters of the Venetian Lagoon. The Italian favourite was out of the race but unknown to him so was de Bernardi, who had also been forced out on lap two by engine failure. Picked up by a patrol boat from HMS *Eagle*, it was discovered that the problem was a loose connecting rod. Adding to the misery of the Italian supporters, it then began to rain which at least took their minds off the poor performance by Guazzetti who was being completely outpaced by the British machines. Was it possible that the British could finish without a mishap and be placed 1-2-3? It certainly looked like it until disaster struck on the sixth lap, a loose spinner forcing Kinkead in the Gloster IV to throttle back and retire from the race in lap five. His third lap speed of 277.18 mph was the fastest biplane seaplane flight ever recorded. To the spectators it seemed that the sixth lap was jinxed as Guazzetti, who was on the point of completing the lap, suffered a fractured fuel line, forcing him to land in the Venetian Lagoon. His retirement from the race was a national tragedy and no doubt responsible for the floods of tears emitted by the patriotic Italian crowd. We also tend to forget that he had a fortunate escape, not only because he was almost blinded by the fractured fuel line, but because as he turned off the course he nearly hit the top of the Excelsior Hotel before landing in the lagoon beyond. Francis 'Rod' Banks, responsible for the S.5 fuel mix, was one of the spectators narrowly missed by Guazzetti's aircraft:

> I went to Venice to see Webster of the RAF High Speed Flight team win and, with the others, including the Crown Prince of Italy, was nearly knocked off the top of the Excelsior Hotel on the Lido when the pilot of one of the Macchis had engine trouble. He managed to turn his machine off the course but appeared to fly directly at us.

Webster is seen standing on the port float above the Supermarine maintenance crew; Mitchell is seen standing in the centre of the line-up. The photograph was probably taken at Calshot after the 1927 contest.

The Armstrong Whitworth Siskin III was the RAF's front-line fighter in 1927 and could only manage a top speed of 186 mph.

The S.5 in which Flight Lieutenant Samuel Kinkead attempted the world speed record on 12 March 1928. He crashed into the sea during the attempt with fatal results.

Banks was writing in the *Journal of the Royal Aeronautical Association* in 1966, but recovered enough composure to witness Sidney Webster winning the race with an average speed of 281.66 mph. Webster not only established a new world record for seaplanes with this win but bettered the existing world record for landplanes by 3 mph. Close behind him was Oswald Worsley in the second S.5 who completed the course with an average of 273.01 mph. It is worth recording that the RAF's front-line fighter in 1927, the Armstrong Whitworth Siskin IIIA, had a top speed of 186 mph!

Back at home Mitchell's success was overshadowed a little by de Bernardi establishing a new world record of 296.94 mph in a Macchi M.52. Not to be outdone, the Air Ministry prepared to recapture the speed record on 12 March 1928 with an S.5 piloted by Flight Lieutenant Samuel Kinkead, the same individual who was forced to retire from the 1927 race in the Gloster IV. The 31-year-old Kinkead was a First World War fighter ace credited with thirty-three victories, a man who began the war with 2 Wing RNAS during the Gallipoli campaign and transferred to the RAF in April 1918. Taking off from Calshot in the late afternoon, the sea mist obscured the passage of the S.5 from those on land and thus precisely what happened is unclear. All that is really known is that the S.5 flew straight into the sea, killing the South African pilot almost immediately. The circumstances of his death have never been satisfactorily explained, although a verdict of death by misadventure was passed at the inquest. The witnesses to the crash thought Kinkead was flying very low and very fast when his S.5 dived into moderately deep water near the Calshot lightship. Although the RAF duty motorboat quickly marked the wreck site, it took a further two days for the salvage vessel to find and retrieve the wreckage that had split into two parts. It was thought that Kinkead had been thrown clear of the machine during the crash but his body was found, minus half of his head, compressed into the tail which had to be cut open in order to retrieve the body. What is perhaps not always appreciated is that the pilots of the Schneider Trophy racers were always about a second away from death and disaster, especially since they usually flew at low altitudes with no seat harness. Kinkead was no exception to this rule. He is buried at All Saints' Church, Fawley in Hampshire.

Flight Lieutenant Samuel Kinkead. Among his medal ribbons can be seen the DSC and bar and the DFC and bar.

* * *

Flight Lieutenant D'Arcy Greig's attempt on the world speed record was not enough for the FAI to proclaim a new record and it remained a British record. This photograph shows Greig in the cockpit of the S.5.

Calling for volunteers from service pilots, Kinkead's replacement was 28-year-old Flight Lieutenant D'Arcy Greig who joined the High Speed Flight on 1 May 1928. Greig, who was admittedly a little dismayed at de Bernardi's raising of the world speed record to 318.57 mph in a modified Macchi 52, was confident of the S.5's performance and went ahead with an attempt on the record in November 1928. However, his speed of 319.57 mph, although heralded as a new world record by many, was not enough to justify a claim to the FAI and became a purely British speed record. For the moment Italy appeared to have the upper hand. However, behind the scenes in 1928 there had been a great deal of movement within the aviation company scene. Vickers (Aviation) Ltd had bought out Supermarine, which became the Supermarine Aviation Works (Vickers) Ltd. The board at Vickers had little doubt as to the potential of the smaller company, especially as the Schneider Trophy designs were reaching their peak. The result of this was the separation of Mitchell's team from the run-of-the-mill Vickers products, which possibly made the design of the Spitfire even more of a certainty.

Earlier that year Jacques Schneider died at Beaulieu-sur-Mer on 1 May 1928 after seemingly recovering from an appendix operation. In those final days he must have considered that the contest to which he had given his name had developed into something so inconsistent with the aircraft he had envisaged using the world's waterways. The aircraft were now lightweight single-seat seaplanes, with an engine life expectancy of just a few hours. What he didn't know, or could not even have been aware of, was that his attempt to advance maritime aviation would ultimately be responsible for a new generation of high-speed fighter aircraft, but that was in the future.

Political manoeuvring had also been at work and at the FAI meeting held in Paris on 5 January 1928, a meeting attended by Jacques Schneider himself, the committee agreed to the trophy being contested biennially. With the matter finally settled, the Royal Aero Club announced preliminary arrangements for the 1929 contest. There had also been a further development in the design of British engines and that was down to the emergence of the Rolls-Royce racing engine. Even though the Napier Lion engine was one of the most reliable – no Schneider aircraft had been forced to retire through engine failure – the question began to surface in the Mitchell team as to whether this engine had reached the end of its potential. It was a view shared by the assistant director for engine development at the Air Ministry, Major George Bulman, who did not think enough extra power could be wrested from the Napier Lion engine to guarantee another victory and consequently he turned to Rolls-Royce. At the time Rolls-Royce had been having considerable success with the Kestrel engine, but Mitchell's only experience of Rolls-Royce engines was with the Sea Eagle flying boat of 1923. After a visit to his home at West Wittering by James Bird of Supermarine and George Bulman, Henry Royce agreed to the development of a new engine, and the concept of the Rolls-Royce R engine was born.

With the re-forming of the High Speed Flight, Squadron Leader Augustus Harry Orlebar was posted to Felixstowe in command of an entirely new group of pilots. In addition to Flight Lieutenant Greig, the other pilots were Flight Lieutenants George Stainforth and Richard 'Dick' Waghorn and Flying Officer Richard 'Batchy' Atcherley, all from the Hendon Aerobatic Team. The 31-year-old Orlebar was well-suited to command, being commissioned as a second lieutenant

Reginald Mitchell pictured with Sir Henry Royce whose engine provided the necessary power for the S.6.

The 1929 High Speed Flight. From left to right: Flying Officer Waghorn, Flying Officer Moon, Flight Lieutenant Greig, Squadron Leader Orlebar, Flight Lieutenant Stainforth and Flying Officer Atcherley.

in the Bedfordshire Regiment in 1915 and sent to Gallipoli. Wounded, he was invalided to the United Kingdom and seconded to the RFC where he trained as a pilot. Posted to 73 Squadron, which was flying Sopwith Camels, he shot down and severely wounded *Leutnant* Lothar von Richthofen near Cambrai. Orlebar was credited with two enemy aircraft destroyed while serving with No. 19 and a further four as a flight commander in No. 73 Squadron. After the war he had also been a test pilot at Martlesham Heath along with Webster.

Despite the Air Ministry proving to be a little reticent, the High Speed Flight lost little time in beginning pilot training on Fairey Flycatchers and Gloster IVs and approaching Supermarine and Gloster to develop new aircraft. It was inevitable that a reliance on the designs of Supermarine and Gloster aircraft for the 1929 contest would take precedence, but Mitchell was faced with considerable difficulties at Supermarine, the principal problem being designing an aircraft for an engine that was new and untried. He knew the proportions, but the power the new engine may or may not produce was still very much in question. The lateness of delivery was firmly placed at the door of Rolls-Royce. In May the engine was producing 1,545hp but failed, and it was only on 27 July that the engine passed the test of one hour at full throttle. As for those civilians who lived within earshot of the Derby

British Domination: 1927–1929   141

The Fairey Flycatcher was used as a training aircraft for the 1929 contest along with the Gloster IV. This photograph depicts a Flycatcher II.

The Supermarine S.6.

works, it must have been a welcome relief when testing finally stopped. The most obvious difference between the S.5 and the new aircraft was the different cowling, made necessary by the V-shape of the new engine and the additional weight of the engine, which required the cockpit to be placed further back. All the same, Mitchell was content to produce an all-metal aircraft, designated the S.6, to take the new and more powerful Rolls-Royce R engine, which was now producing 1,900hp. It was a larger aircraft than the S.5 and to accommodate the thirst of the new engine, the floats both contained fuel tanks and were slightly larger to provide the additional buoyancy required. The new engine was going to require something like 2.5 gallons each minute! The wings were braced by streamlined wires similar to those of the S.5, and to provide extra liquid cooling, each float was equipped with a surface radiator.

With all this frantic preparation it was not until Monday, 5 August that the first S.6 was ready to be launched into the water at Woolston and towed round to Calshot, where elation turned to despair as it was found that the greatly induced torque from the engine had not been fully anticipated. The S.6 had a tendency to dig in to the surface of the water with the left float. Richard Waghorn described the first flight of the S.6 in the *Journal of the Royal Aeronautical Society*:

> With the arrival of the S.6 our hopes had risen considerably, only to be immediately lowered to the depths when Squadron Leader Orlebar started his initial tests in Southampton Water. The S.5 in her take-off had been so straightforward that we assumed her elder brother would also prove himself equally docile while being broken in. We were therefore very surprised to see the behaviour of the S.6 on her first test. The S.6 behaved much as a horse refusing a fence. She sat on her tail and it seemed as if no amount of coaxing would get her forward. Furthermore, she dug her left wing into the water and, not content with so much mischief, started a gigantic porpoising.

Failing to even get airborne, Orlebar eventually gave up the struggle and returned to base. Mitchell's response to the increased torque was to lengthen the starboard float to carry 90 gallons of fuel and reduce the capacity of the opposite float to 25 gallons; all the problems were capable of resolution and further tinkering with the cooling system resulted in wing scoops being fitted to allow the radiators to increase their efficiency. When Orlebar made his next attempt, he decided to employ the old seaplane trick of putting firm backward pressure on the stick when flying speed was reached and hauling the machine off the water. It was successful, and from then on the High Speed pilots employed this technique. However, the S.6 was still a difficult aircraft to fly and these high speeds achieved by the Schneider machines required good visibility, gentle winds and choppy water. Too much of a swell would cause the front of the floats to dig in while flat calms would prevent the S.6 from 'unsticking' from the surface of the water. The technique that finally evolved was described in detail by Richard Waghorn:

Richard Waghorn climbing into the cockpit of the S.6.

> We found it essential to point the machine about 70 degrees to the right of wind and to have right rudder on from the start. The machine then runs along with its left wing a few inches from the water, across wind but not swinging. She is clear of the spray which, up to 30 mph completely envelopes the pilot. Having got her, therefore, running across wind at 40–50 mph one is now confronted with what is really the trickiest part of the proceedings, and that is to get her into wind without letting her swing right round, which she will want to do; once left rudder is applied the machine will accelerate rapidly and provided you have not put on too much rudder should reach her hump speed by the time she is directly into wind. At this point she assumes a new position on the water very much lower in front, and accelerates rapidly up to taking-off speed.

Cornering technique was another matter of great importance, and in 1929 especially, one of the main questions was whether the turn should be tight or loose. At 300 mph the most efficient turn lay somewhere between 4 and 6G but, for example, with a tight turn the pilot lost more speed and travelled a shorter distance. In an effort to apply a little more science into the debate, several of the training aircraft were fitted with instruments measuring acceleration, speed and climb. The data obtained indicated that the perfect turn was in fact a vertical one, performed as level as possible with the stick pulled back. Certainly the pilots found that they stood less of a chance of blacking out if the aircraft was brought gradually and evenly into the turn.

In the meantime, Henry Folland was going ahead with the design of the Gloster V from which the Gloster VI evolved. A departure from the biplane format, this

A frontal view of the Gloster VI.

monoplane boasted a Napier Lion engine producing 1,320hp and was considered to be smaller and lighter than the S.6 and, what's more, could possibly challenge it for speed. The aircraft was a favourite of George Stainforth's, who preferred the Gloster to the S.6. Yet while the aircraft showed promise and high speed, it had problems with fuel supply when banking, which led to engine cut-outs on the turns and even in level flight. Alarmingly, the faults with the Gloster's fuel system persisted, which appeared to defy any correction despite the ground crew working day and night to locate the problem. The night before the race both Glosters were withdrawn, leaving Greig to lament that 'had Folland designed the Gloster VI for the Rolls engine instead of the Napier, I believe it would have walked away with the contest.'

The defeat of Italy in 1927 after their victory at Hampton Roads was followed by a change and reorganization that was largely due to the efforts of General Italo Balbo, the Italian airman and Fascist leader who played a decisive role in developing Benito Mussolini's Air Force, a man who we first met in Chapter 4. This meant widening the net to take in all the leading aircraft manufacturers, the creation of a High Speed Flight and moving their previous base from Lake Varese to Desenzano on Lake Garda. Accordingly, aircraft were ordered from Macchi, Fiat, Piaggio and Savoia-Marchetti and engines from Fiat and Isotta Fraschini. In addition to pilot training, the large expanse of water offered sheltered facilities akin to the open nature of the sea as well as research into high-speed flight. Of the four new Italian designs, all were all monoplanes and very different from each other with Macchi being chosen without hesitation. The M.67 designed by Castoldi was a direct descendant of the M.39 and M.52 and constructed largely of wood with metal forward of the cockpit to ensure a rigid mounting for the wing, engine and floats. Powered by an eighteen-cylinder, 1,800hp Isotta Fraschini engine developing 1,400hp, the M.67 was very much like its predecessor but, like Mitchell, Castoldi had to alter the design to accommodate the heavier engine. The choice of engine

General Italo Balbo played a decisive role in developing the Italian Air Force.

The Macchi M.67 was a direct descendant of the M.39 and M.52.

The Fiat C.29.

The Savoia-Marchetti S.65.

was dictated by much the same reasons for which Mitchell had chosen the Rolls-Royce: the Fiat AS.3 engine was thought to have reached its full potential. Fuel for the aircraft was carried in the floats and additional surface radiators were situated between the float struts. The other distinguishing feature of the M.67 was the use of a three-bladed propeller.

The Piaggio-Pegna P.c.7 was described as a hybrid resulting from the crossing of an aeroplane with a submarine.

Fiat produced the C.29, the smallest of the new machines, which was built around the latest Fiat engine, the 1,000hp AS.5 engine, said at the time to be the lightest in the world. It was similar in many ways to the Macchi designs and at first glance only the tail unit, which was almost in line with the upper surface of the fuselage, appeared to be different. The airframe was all-metal and although its wingspan was a mere 22ft, the wing loading was kept low. A month before the trophy race Francesco Agello was test-flying the C.29 when it dipped forward and set down heavily on the water surface, wiping off the floats as it did so. Agello miraculously escaped death and survived with the addition of a few scratches. Seeing there was no injury, General Balbo ordered the trials to continue, but on 22 August the 35-year-old Captain Giuseppe Motta was killed just after his M.52 had reached a speed of 362 mph. His machine dived steeply into the lake and Motta was killed on impact. Quite why the accident occurred is unclear, but one report centred on a stall while another suggested that dense smoke fogged the windscreen. Whatever the reason for the smoke, Macchi fitted ventilating tubes to protect future pilots from the danger of noxious fumes.

The other two designs were very different from those expected. The Savoia-Marchetti S.65 was admittedly a floatplane, but it had twin booms supporting the tail unit which allowed two Isotta Fraschini 1,000hp engines to be mounted in a central nacelle. It was a revolutionary design and the contra-rotating propellers were expected to provide formidable power and eliminate all torque problems on take-off. The Piaggio-Pegna P.c.7 was the fruit of many years of research by Giovanni Pegna which, rather unkindly, has been described as 'a hybrid resulting from the crossing of an aeroplane with a submarine.' It had neither floats nor a conventional flying boat hull and was constructed with a watertight fuselage with a marine propeller to allow it to get up on its hydrofoils, at which point a conventional propeller would be engaged for take-off. Interestingly it had a cantilevered elliptical wing, both of which would be used by Mitchell in the designs leading to the Spitfire.

The delays experienced by the British were duplicated in France and America. Despite the two seaplanes built by the French Bernard Company, the HV.40 and HV.42, making their maiden flights, the engines were a different matter. France had not competed for six years and the tight schedule began to show. The Gnome-Rhône Mistral was unlikely to be ready in time and the Hispano-Suiza was behind schedule and to fill the gap in the schedule the Hispano-Suiza V-type engine was used to make pilot training possible which took place in the interim. The Nieuport-Delage was even more behind and there was little hope it would even have been ready for the navigability trials. The lakeside military seaplane base at Hourtin in the Gironde was set aside and the pilots selected for testing were Sadi-Lecointe for the Nieuport and Florentin Bonnet for the Bernard machines. Bonnet still held the

The Bernard Ferbois V.2 in which Florentin Bonnet broke the landplane speed record. Note the monoplane configuration.

The Mercury-Williams racer, looking like a cross between the Macchi M.67 and the Gloster VI.

world speed record of 278.37 mph for landplanes flying a Bernard-Ferbois V.2, but on 6 August Bonnet was attempting a loop in a Nieuport 62, which had been used for high-speed training, and crashed the aircraft, killing himself. Within forty-eight hours of the crash France withdrew, citing inadequate pilot training. We shall probably never know the real reason for the French withdrawal, but the crash must have had something to do with it.

The private American entry of the Mercury racer financed by the Mercury Flying Corporation and piloted by the indefatigable Al Williams was also running into trouble and by mid-August he had got no further than his preliminary taxiing tests. With the constant failure to get airborne, Williams finally realized that the aircraft was in fact too heavy and a new Packard engine was required. Running out of time, Williams was forced to withdraw.

As for the Italians, neither of their machines were ready and although work continued, their only hope of an effective challenge to the British lay in a thirty-day postponement. The Italian air attaché formally requested such a postponement, giving bad weather and the loss of Giuseppe Motta as their reasons. The Italian request was denied by the organizing committee, basing their judgement on their interpretation of the rules and under these rules the date of the contest had been fixed and they had no power to change them. The British point of view was reasonable in the circumstances: if Italy was to provide a challenge, then a delay of another few weeks in which to get ready may well alter the outcome of the race.

Naturally the Italian press was vociferous on the subject and urged the Italian team to withdraw completely, but General Balbo was unmoved and announced on 28 August that the Italian team would take part in the race:

> The probability of victory by our team, already very hard pressed by the very great delay of the building firms in delivering the machines and the engines, is almost annulled by the death of Captain Motta. His death not only deprives us of our best pilot, but has cost us the engine and the machine which had been perfected for the best speed, and has also meant a week's interruption in our programme.

Any hopes that the Italians would be a no-show, in which case the British would fly unopposed round the course, were dashed by Balbo's announcement. Instead the team left for Calshot, arriving on 29 August with the two remaining M.67s, a veteran M.52R from the previous race, as well as the second Fiat C.29 and the Savoia-Marchetti S.65.

Three of the Italian team at Cowes in 1929. Left to right: Sergeant Major Franceso Agello, Captain Alberto Canaveri and Warrant Officer Tommaso Dal Molin. Of these three, only Tommaso Dal Molin in the M.52R flew in the contest.

The course at Calshot consisted of seven laps of a quadrilateral 30-mile circuit in the Solent. With the start and finish line just off Ryde Pier, the turning-points were a pylon off Seaview leading to a destroyer-mounted pylon off Hayling Island, a pylon off Southsea pier and a destroyer-mounted pylon off West Cowes, which left the final stretch of 6.28 miles back to Ryde Pier.

On 29 August the Italians arrived at Calshot under the leadership of Lieutenant Colonel Mario Bernasconi, but it was not until nearly a week later that Lieutenant Giovanni Monti was able to fly the first of the M.67s, the same day that Warrant Officer Tommaso Dal Molin had the misfortune to hole a float in the M.52R. The navigability tests passed without incident for both teams except that Atcherley's S.6 appeared to develop a decided list which fortunately was repaired the next morning. What was more devastating was the white fleck of metal detected in Waghorn's aircraft during a final check. It was again very fortunate that a team of Rolls-Royce fitters from Derby had come down to watch the race and those that were sober enough were rapidly drafted in by Cyril Lovesey, the Rolls-Royce engine specialist, to change the engine block. Such was their knowledge of the engine that by the time Orlebar put in an appearance the following morning, the job had been completed. Incidentally, one of the group who worked on the engine was Jack Warwick, who went on to become manager of the experimental department in Derby.

Race day was as near perfect as the Royal Aero Club and the organizers could have wished, and hundreds of thousands of spectators lined the Hampshire shores and the Isle of Wight along the Solent. Everyone of that army of spectators that had scattered itself along the promenades and beaches expected to see aeroplanes flying faster than anything that had previously taken to the air. In that they were not to be disappointed.

The British entries from Supermarine were resplendent in their silver and blue and Italy's Macchis were painted in bright red, the *corso rosso* from Italy's racing tradition. Among the knowledgeable crowd many recognized that the M.67 was a match for the Supermarine S.6 with its 1,900hp Rolls-Royce R engine, but only if its engine would hold together long enough to complete the course. The first aircraft away was Waghorn's, and using the take-off technique they had developed in practice, was soon heading for the pylon at Seaview. Although his first lap was a little disappointing, he soon settled down, clocking a steady 329 mph. The next aircraft away was the M.52R flown by Tommaso Dal Molin followed by Greig in the S.5. The M.67 flown by Lieutenant Remo Cadringher was the second Italian entry into the air, but was restricted to using below maximum engine revolutions to avoid engine failure and, almost blinded by exhaust fumes that whipped into the cockpit, he landed before he was totally incapacitated. Atcherley in an S.6 was suffering vision difficulties of a different nature to Cadringher in that the slipstream

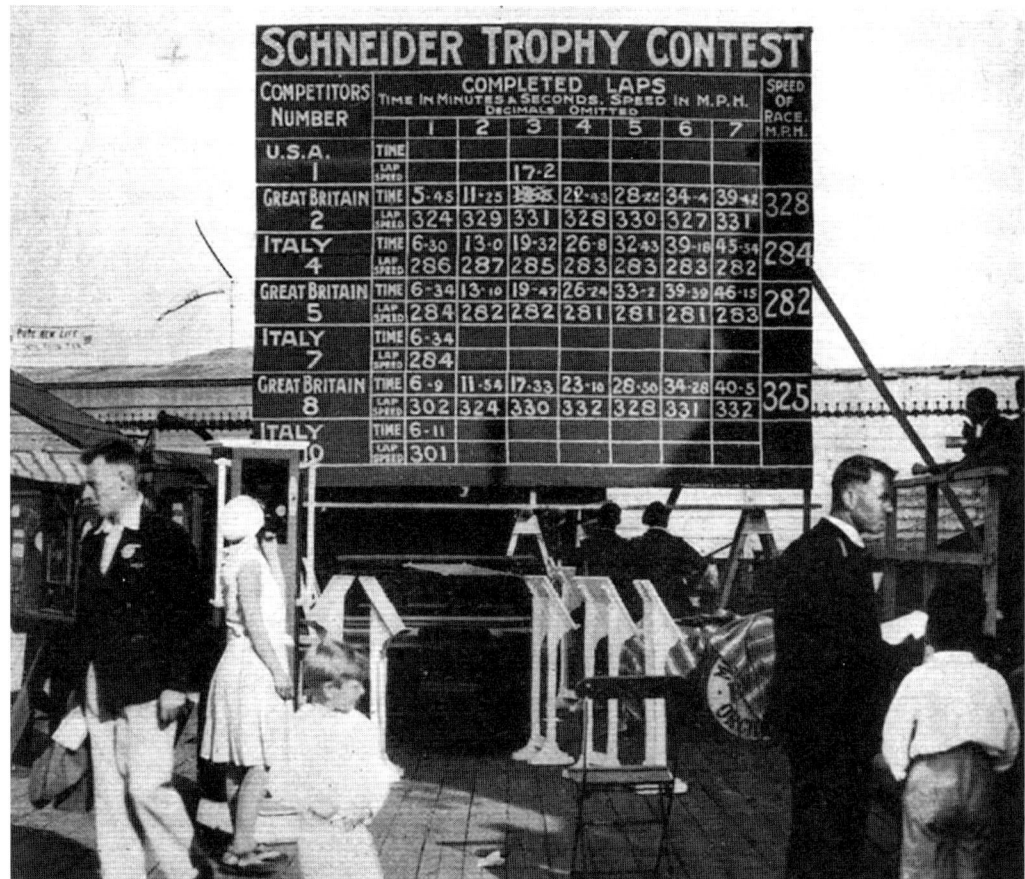

The 1929 Schneider Trophy Race scoreboard erected on the seafront for the benefit of the spectators.

had knocked his goggles off-centre and prevented his spare pair from being put into place with only one hand. Tucking down low in the cockpit, he approached the Seaview turn. Major Alan Goodfellow, a member of the racing committee of the Royal Aero Club, was in charge of the Seaview turning-point:

> Batchy came near to killing me – nearer I think than any German ever did – when I was at the Seaview turning-point, which consisted of an old destroyer on which a pylon some 30ft high had been mounted. As the race was between the British and Italian teams, we had two observers of each nationality. In order that there should be no possibility of mistakes I decided that I and one of the Italian observers should climb to the top of the pylon, leaving the other two on deck. Dick Atcherley duly took off in his Supermarine S.6 and headed in our direction. As he came rapidly nearer we realized he was heading straight for the top of the pylon.... At the very last moment he saw us and swerved sharply, passing no more than a wing span inside the pylon.

Even though Atcherley was disqualified, he set up the fastest lap of the entire race and completed the course, only to be confronted by Orlebar and told the bad news. Behind Atcherley was Italy's last hope, Lieutenant Giovanni Monti flying the second M.67, but success was not to be his. Almost immediately after completing his first lap a serious leak in the engine cooling system sprayed him with boiling water and steam, forcing him to put down in the emergency landing lane beyond Seaview. Badly scalded, he completed his trophy race in hospital. The Italian challenge was over.

For Waghorn, all was going according to plan. The air was hot in the cockpit, but a continual stream of cool air from the ventilating pipe was cooling his face, the engine temperature was steady and he was holding a height of 200ft above the water. However, like Webster in Venice, he too had become confused about how many laps of the course he had in fact flown. Suddenly the misfiring of the engine near Hayling Island gave him a horrible fright: seconds later the engine began missing badly, providing him with but one conclusion, fuel starvation! Pulling up to 800ft, he hoped beyond hope that if the engine cut out completely, he would still be able to glide to the finish line. As he approached the Cowes turn he realized that his chances of a glide to the finish were dashed and he was forced to put down off Old Castle Point. His race was over, or so he thought! Failing to understand why the individuals in the motor launch that had come alongside him were so excited, he finally comprehended that he had been flying an extra lap and his aircraft had run out of fuel. Waghorn's own account was far more lucid:

> The rate that petrol can be poured out of a 2-gallon tin will give some idea of the rate the engine was consuming its petrol during the race. I had therefore been told that on no account to use full throttle as I shouldn't finish the course. Imagine then my feelings when the engine momentarily cut right out and started missing badly just after what I imagined was my sixth lap. Would the Rolls engineers ever believe that I hadn't given full throttle? I began to gain height and continued round the course with the engine spluttering and only taking about half throttle…. It was twenty minutes later that I learnt I had done an extra lap, and I also realised how deadly accurate had been Lovesey's estimation of the petrol consumption.

Thus Britain retained her hold on the Schneider Trophy with a Supermarine S.6 flown by Waghorn. He took the trophy with a race speed of 328.63 mph, about double the top speed of the front-line fighters of the day, with Dal Molin second with 284.2 mph and Greig in third place with 282.11 mph. The 25-year-old Richard Waghorn had but two years left to live. On 5 May 1931 he was test-flying a Hawker

Richard Waghorn was killed in May 1931 while test-flying a Hawker Horsley biplane. This photograph is of a Horsley II which played an important part in the development of the Merlin engine.

Horsley biplane from Farnborough when he lost control in high winds. He and his passenger both parachuted from the aircraft and while the passenger landed on the roof of a factory sustaining only minor injuries, Waghorn was seriously injured and died on 7 May.

The 1929 contest had been a close-run thing in that had the experimental team of Rolls engineers not arrived when they did to change the block in Waghorn's machine rather than driving down on the morning of the race, it is likely that the Macchi 52.R flown by Warrant Officer Tommaso Dal Molin would have won the race. For the Italians, the victory that had nearly been won and the disappointing effect of months of effort was plain to hear in Mario Bernasconi's reaction to the question of whether Italy would compete the next time. 'I don't know if we want a next time; the cost of taking part in this contest is simply staggering.' General Balbo's reaction was more upbeat: 'We have obtained the results we expected, but we have now finished playing our part as sportsmen. Tomorrow our work as competitors begins.'

Prime Minister Ramsay MacDonald, responding to insistent demands for a comment on the race, was equally buoyant, stating that 'We are going to do our level best to win again.' It was a comment that he would live to regret making.

Chapter Eight

The Final Flourish: 1931

'The contest was bad from the service viewpoint, bad from the viewpoint of efficiency, and bad for RAF morale as a whole.'
                    Marshal of the Royal Air Force, Sir Hugh Trenchard.

The flying undertaken by the Schneider Trophy pilots was hardly normal flying; it was a highly-specialized form of test-flying for which there had been no precedent in the history of aviation. If Waghorn's report, for example, of the take-off methods employed by the S.6 had been written up by a normal test pilot it would have reflected badly on the designer but, as we know, with the Schneider racers, normal standards did not apply. By taking the first steps into the unknown world of high speed, the Schneider pilots plotted the course that others would follow but sadly, there was a human cost. In the preparations for the 1931 contest no fewer than five pilots died from aircraft crashes and another from peritonitis, which may or may not have been brought on by the stresses of high-speed flight.

In Britain the anticipation of another Schneider Trophy win that would secure the trophy was severely deflated by the news that neither the government nor the Air Council wanted anything further to do with the contest. This point of view was exacerbated by Hugh Trenchard's position in that he could see little of value in the continuation of the contest. Having pressed for RAF participation in 1927, Trenchard was now proposing that the government withdraw on the

Marshal of the Royal Air Force, Sir Hugh Trenchard.

basis that expenditure was no longer justified by the rewards, which was exactly the reason that the Americans gave for withdrawing from the contest after their defeat in 1926. To an extent one can sympathize with the British government: there was a growing feeling in both Britain and Italy that the burden of even a biennial contest was becoming too much, and added to the financial crisis in America the government's action did look logical. The final decision was taken in the Cabinet meeting of 25 September followed by an Air Ministry statement that the RAF would not be supporting the contest and all future participation would be down to private enterprise.

With the news of private enterprise ringing in their ears, Supermarine and Rolls-Royce wrote to the Air Ministry predicting an increase of 25 mph on the understanding that the two existing S.6s would be loaned back to them and the pilots of the High Speed Flight would be employed to fly them. Their request met with a flat refusal by the government to lend either the machines or pilots, the Air Council fearing that participation would involve them in further expense. It was an unbelievable situation that many considered to be inconsistent with the general mood of the country. Ramsay MacDonald, who you will remember said in 1929 that 'We are going to do our level best to win again', received a deputation that included the chairman of the Royal Aero Club where it was pointed out that participation would not exceed £100,000 and the club had already received £22,000. MacDonald had already been stung by the constant attacks in the British press and promised to reconsider the matter. The Royal Aero Club meanwhile explored every avenue open to them and left no stone unturned in their quest to force the government into supporting the 1931 contest. Charles Grey, the prominent aviation writer, made the remark that 'a government that will give £80,000 to subsidise a lot of squealing foreigners at Covent Garden and will refuse £80,000 to win the world's greatest advertisement for British aircraft is unworthy of the Nation.' The *Stoke Evening Sentinel* ran an article reflecting the general unhappiness of the aviation industry and particularly among the workforce at Supermarine. In an interview with Mitchell the paper is recorded as saying that British aircraft

Lady 'Fanny' Lucy Houston.

were unquestionably superior to other aircraft in the world, but if the research work was dropped at that time, that position could be lost.

Fortunately for MacDonald's Labour government, their face was saved by the actions of a rather formidable, public-spirited and patriotic woman who had led an interesting life and rose to become the second-richest woman in Britain. Lady Lucy Houston, as she became after her third marriage, at the age of 16 became the mistress of a wealthy brewer called Frederick Gretton, who left her £6,000 a year for life when he died. She married Theodore Brinckman in 1883, but was divorced in 1895. After her second marriage in 1901 to Lord Byron, a descendant of the poet, she became an active suffragette. Widowed in 1917, in 1924 she married the shipping magnate Sir Robert Houston who died in 1926, leaving her £5.5 million. Approached by Colonel the Master of Sempill, she agreed to sponsor the 1931 British Schneider Trophy team to the tune of £100,000 and at the same time embarrass the government. Her press release revealed her dislike of MacDonald and his socialist government:

> I am utterly weary of the lie-down-and-kick-me attitude of the Socialist Government. To plead poverty as a reason for objecting to England entering a race against teams supplied by nations much less wealthy than our own is a very poor excuse. To down anything that extols and glorifies the wonderful spirit, that even a Labour Government cannot knock out of us British, seems [to be] their chief aim. It is down with the navy, down with the army, down with the Air Force, down with our supremacy in India – but up with Gandhi, up with strikes which every honest workman detests, the ultimate aim of which is to bring about revolution and ruin and beggary of all in the kingdom.

Her cheque lost little time in being sent to the secretary of the Royal Aero Club; in reality an unfortunate occurrence for the government but fortunate for Supermarine and Rolls-Royce in that the means were now guaranteed for Mitchell to produce a third and final design in the quest to win the Schneider Trophy. To the government's credit, they took a great deal of abuse without ever seeking to blame or criticize their air advisors, especially as none of these advisors, Trenchard included, ever rose to defend the government.

On 22 October 1929 a meeting was held with the backing of the Italian government where plans were discussed with prominent Italian engineers about the new machines for the 1931 contest. A similar meeting called by General Balbo into the failures of the 1929 contest concluded that the distribution of work among so many different companies had been a mistake and consequently only Macchis and Fiat would be considered in future. Of the four differing machines prepared for the 1929 contest, further development of the Piaggio-Pegna was abandoned, as was the C.29. The Savoia-Marchetti 65 came to a sticky end in January 1930 when, during an attempt on the world speed record, it crashed into Lake Garda, killing its pilot, 28-year-old Tommaso Dal Molin, who was reportedly overcome by fumes from the engine. The next Italian pilot to be killed was Giovanni Monti on 2 August flying a Macchi M.72 in an accident that threw the Italian plans for moving to Britain into total disarray. The Macchi MC.72 was essentially a twin-engine aircraft, with two of the 1,000hp Fiat AS.5 racing engines coupled together in tandem, one behind the other, driving counter-rotating propellers through a central co-axial drive shaft. The final specification was called the Fiat AS.6 and, with modifications, it produced an extraordinary 2,850hp. Captain Monti was throttling down after take-off and aiming to overfly Desenzano to demonstrate the misfiring of the engine to the ground crew when his aircraft spun and dived into Lake Garda. Monti's body was not recovered by fishermen until nearly a month later.

On the morning of 3 September the Italian Air Attaché in London called on the secretary of the Royal Aero Club and informed him that unless a postponement of at least six months was granted, Italy would have no option but to withdraw, and

Francesco Agello seated on the float of the M.C.72. He set a new Absolute World Speed Record of 423.82 mph in 1933 and in 1934, again flying an MC.72, raised it to 440.68 mph. That record still stands today.

a few minutes later the French Air Attaché made a similar call. Both deputations cited poor weather and the loss of pilots and aircraft as the principal reasons for their call and dwelt on the unsatisfactory nature of a fly-over for the defending nation. The Royal Aero Club replied almost immediately, stating that the rules did not allow for any postponement except on a day-to-day basis and consequently would be carrying out its programme regardless. On 5 September both Italy and France withdrew from the 1931 contest.

Without doubt both delegations must have reminded the Royal Aero Club of a similar occurrence that took place in 1924 when the Americans agreed to postpone the contest for twelve months. The Royal Aero Club had expressed their appreciation at the time and no recourse to the rules had been made. However, in 1931 things were a little different in the British team. First of all they only needed one more victory to secure the trophy and it looked very much as if this was about to happen; second, it looked as if this was Britain's last throw of the dice, which was manifested by a deep-seated desire not to let down Lady Houston's sponsorship; and third, all the elaborate preparations had been made at Calshot and a very large expenditure had already been incurred by all concerned. Quite naturally, insinuations of a lack of sportsmanship from France and Italy abounded, leaving the British with little alternative but to announce that even if there was no opposition, the race organization would continue and Britain would complete a fly-over of the course if necessary.

The reader will recall that in France the situation was a little different. After their failure to enter for the 1929 contest, the French realized that performance requirements had moved on significantly and the gap in practical experience and expertise was going to be difficult to make up, particularly after years of non-participation in the contest. Orders were placed with the Bernard and Nieuport companies as well as the Dewoitine company for low-winged, wire-braced monoplanes and the 1929 machines, which were described in Chapter 7, would now be used for training. Tragedy struck early with the death from peritonitis on 15 June of the 34-year-old test pilot Antoine Paillard, a First World War flying ace credited with five aerial victories. Although dreadfully missed, Paillard was eventually replaced and by July most of the pilots had flown one of the Bernard 120s. However, greater misfortune was just around the corner. Fernand Lasne crashed into the Seine while taking the ND.450 back to the makers for an overhaul; fortunately he escaped without injury, which was not the case on 24 July when Lieutenant de Vaisseau Bougault, flying a geared Bernard HV.120, crashed into the lagoon at *Étang de Berre* and was killed. His death was a particularly severe blow as he was the only service pilot competent to fly the Bernard HV.120, geared or ungeared. The cause of the accident has never been satisfactorily determined,

The Bernard HV.120. It was an aircraft such as this that killed Lieutenant de Vaisseau Bougault.

but there had been problems with the airscrew and perhaps the propeller breaking up in flight caused the aircraft to dive uncontrollably into the water. Technical investigation of the wreck later discovered that a spark plug was missing and this maybe penetrated the windscreen and caused the pilot to lose control. Whatever the cause, Bougault was dead and France, as we know, withdrew from the contest on 3 September.

Incidentally, throughout 1931 it was rumoured that the 40-year-old Sadi-Lecointe, the pilot who had won the final Gordon Bennett Cup for France in 1920, would fly the Nieuport-Delage, and the well-known French aviator Jean Assollant, who was credited with two kills in the First World War, would fly the second Bernard HV.120 in the Schneider contest. It is perhaps consistent with the story of the French entry that despite the long training of the service pilots at the *Étang de Berre*, two civilian pilots were chosen to fly the French entries at Calshot. Jean Assollant was killed in May 1942.

There was yet one more tragedy to strike the Italian team, two days before the race was to begin. Seven days after the date on which the Italian Air Attaché had requested a postponement, Lieutenant Stanislao Bellini was killed on the far side of Lake Garda. Italy was clearly intending to upstage an inevitable win by Britain by going for a new world record and after a few successful runs during which the MC.72 was pushed to 394 mph, he was seen to fly directly into a hillside. Bellini was killed on impact and afterwards pieces of the aircraft were found along the flight path leading to the crash site, indicating that an explosion had occurred.

The RAF/RN High Speed Flight 1931. From left to right: Flight Lieutenant Eustace Hope, Lieutenant Gerry Brinton RN, George Stainforth, Flight Lieutenant Francis Long, Squadron Leader Augustus Orlebar, Flight Lieutenant John Boothman, Flying Officer Leonard Snaith and Flight Lieutenant W.F. Dry (Engineering Officer).

It was not until the last day in January 1931 that Supermarine and Rolls-Royce began to think about the impending contest. Government and Air Ministry refusal to fund the 1931 contest had delayed proceedings somewhat, but at least they were now cleared to begin preparations. Flying Officer Leonard Snaith was added to the High Speed Flight which now consisted of Flight Lieutenants John Boothman, Eustace 'Freddy' Hope and Francis Long. Atcherley and Greig had been posted overseas and Waghorn was apparently not even considered! Orlebar was chosen to lead the team again and to his delight Flight Lieutenant George Stainforth was retained for a second year.

Back at Supermarine it was rapidly becoming apparent that more power from the engine was required, but in the seven months remaining before the next race it was an impossible dream. Yet Rolls-Royce engineers were sufficiently satisfied that the R engine could be altered to produce in excess of 2,000hp, although some considerable modifications would have to be made. However, by 12 August the engine was test-run for an hour without failure, producing 2,350hp. Like Rolls-Royce, Supermarine and Mitchell felt that it was impossible to design a totally new

The second S.6B in which George Stainforth flew a new world speed record of 379.05 mph.

The S.6B in which John Boothman won the final Schneider Trophy contest with an average speed of 349.08 mph.

aircraft; consequently they modified the existing S.6 to take the more powerful engine promised by Rolls-Royce. The S.6B had the same basic dimensions as the S.6 of 1929 except it was slightly longer because of the increased length of the floats; the observant will also notice that the tailplane structure was slightly smaller. The increased fuel consumption meant that the floats had to be enlarged to accommodate more fuel and to take into consideration a change in the rules. In order for the whole contest to be held in one day and to avoid bad weather causing a postponement during the navigability tests, competing aircraft were required to take off and land immediately prior to the race instead of the day before. Thus floats would now have to support the weight of the increased fuel as well as that required for flying the circuit. The two S.6s from 1929 were modified to accept the new floats and the latest Rolls-Royce engines and were redesignated S.6As.

Flying with the new machines was a tricky business. The first setback occurred with an accident when Eustace Hope practically wrote off one of the S.6A machines when a piece of cowling worked loose in flight. Attempting to land, the wash of a passing ship caused the aircraft to cartwheel and sink. Eustace Hope managed to survive, but a ruptured eardrum was enough for him to be replaced by Lieutenant Gerald 'Gerry' Brinton, a Royal Navy pilot who had been released from the High Speed Flight and was now recalled. On 18 August the 24-year-old Gerry Brinton, the baby of the High Speed Flight, found himself sitting in a S.6B and after a careful explanation of the quirks of the machine by Orlebar, was seen to make a short run before positioning himself for take-off. Then, having gained a height of about 10ft above the water surface, the tail came up and the aircraft hit the water. Brinton had obviously instinctively eased the stick forward when the aircraft became airborne and not eased it back as Orlebar had noted in Chapter 7. After hitting the water the first time, the aircraft bounced back into the air at an even steeper angle before plunging nose-down beneath the waves, tearing off the floats in a dive that broke Brinton's neck. Sadly, the ravages of the Second World War also claimed the life of Eustace Hope when he was killed in action at the beginning of August 1941 flying an 87 Squadron Hurricane.

Brinton's death heralded a period of inclement weather. The S.6A, which had been crashed by Eustace Hope, had been reconditioned and flew on 6 September, the same day as the two S.6Bs, both of which had been modified to bring the centre of gravity forward. Gone was the instability which had been apparent on the turns and there was a great improvement on take-off, reinforced no doubt by Orlebar who was continually emphasizing the need to keep the stick well back until the nose was 'hanging on the propeller'. No-one had any more trouble with the longitudinal stability at the point of take-off after then, so Gerry Brinton's death had at least served some purpose!

The course for the 1931 contest was triangular and, as before, started and finished off Ryde Pier before it led down past Seaview to a pylon mounted on a destroyer off St Helen's Point. Turning there, the next leg extended right across the Solent to a second pylon situated on the coast at West Wittering. An acute turn took the course past Hayling Island and Portsmouth to another pylon mounted on a destroyer before returning to Ryde Pier.

The plan outlined for the fly-over was simple enough: one of the S.6Bs would fly round the course and attempt to beat the existing Schneider Trophy record as well as the 100-kilometre record. If the first machine failed to do so, an attempt would be made by the surviving S.6A, followed by the second S.6B. The weakness of the plan, as far as the spectators were concerned, was obvious: if the first S.6B completed the course successfully then neither of the other aircraft would fly. Something was required to inject a little excitement into the proceedings and short of a race between all three aircraft the best plan appeared to be an attempt on the absolute world speed record.

With the realization that only two pilots would be required to fly on race day, Orlebar was left in a dilemma: who to choose? Eventually it came down to a question of seniority within the flight. Stainforth, although senior, elected to attempt the world record, while to John Boothman fell the responsibility of completing the course and finally bringing the trophy home to Britain. This meant that Stainforth

Flight Lieutenant John Boothman.

Flight Lieutenant George Stainforth, the first man to fly at over 400 mph.

would fly in the second S.6B and in case he was needed, Snaith would fly the S.6A. Conscious that the eyes of the world were upon them, the team gathered just after 1.00 pm to see a somewhat nervy John Boothman and his aircraft slip into the water. Taking off without any difficulty and landing again almost immediately, as required by the new rules, he was away again and heading for the St Helen's turn. For each of the seven laps he flew with the engine slightly throttled back and all within 4 mph of each other to finish with an average speed of 340.08 mph. Boothman had been airborne for forty-seven minutes and as he passed the finishing line the entire shipping lane broke into a whistling and hooting which almost drowned out the cheering of the crowd assembled at Ryde.

Then, according to the plan and as if to emphasize the superiority of the S.6B, Stainforth proceeded to break the absolute world air speed record with a superb performance, concluding with a maximum speed of 379.05 mph. Yet that was not enough for Stainforth and Mitchell. On 29 September he made a perfect record run in an S.6B over 4 timed miles in opposing directions and achieved an average of 407.5 mph, being the first man in the world to exceed 400 mph, a speed that was not to be achieved by an operational fighter until eleven years later. For this achievement he was awarded the Air Force Cross on 9 October 1931.

Tragically, Stainforth was another officer who was caught up in the ravages of war. In October 1941 he was appointed Officer Commanding 89 (Night Fighter)

Lady Lucy Houston photographed aboard her yacht with the RAF High Speed Flight. Reginald Mitchell is standing on the right and John Boothman is sitting to the left of Lady Houston.

Squadron and on 27 September 1942 he was killed in action flying a Beaufighter at Gharib, near the Gulf of Suez. Today the Stainforth Trophy is awarded annually by the Air Officer Commanding RAF Strike Command to the operational station within the command which has produced the best overall performance in the preceding year and has been most effective in the delivery of support to operations, the development of its people and its readiness to respond and adapt. The station must also demonstrate that it has enhanced the Royal Air Force's reputation and core values. In 2019 the trophy was awarded to RAF Wittering.

Mitchell's success with the S.6B was celebrated on board Lady Houston's steam yacht with a celebratory lunch which was also attended by his wife and the High Speed Flight. As a self-effacing man who shunned publicity, Mitchell was a little horrified to find that he was listed in the official Honours List by George V and became a Commander of the Order of the British Empire.

Nevertheless, the fly-over at Spithead was regarded by the other nations as opportunistic and left a bad taste in the mouths of many. Mitchell himself must have considered the possibility of the Italians winning the trophy in 1931, although given the state of Macchi, to make a successful challenge it would appear that this was perhaps a non-starter. Even so, one is left wondering, especially after Francesco Agello, a pilot who had been part of the Italian team in 1933, set a new absolute world speed record of 423.82 mph in an MC.72 and eighteen months later raised it to 440.68 mph. That record stands even today, almost ninety years later, as the fastest speed ever attained by a propeller-driven seaplane; true testimony to the amazing engineering talent of Italy's design team and its chief speed designer Mario Castoldi. After this success, the MC.72 was never flown again.

Chapter Nine

A Star is Born

'On 3 June 1936, only seven days after it had been delivered [to Martlesham Heath] and when it had done a minimum of official test flying, and before any formal report had been submitted to the Air Ministry, a production contract for 310 aircraft was placed with Supermarine.'

Jeffrey Quill, writing in *Birth of a Legend: The Spitfire*.

The Type 224 design was Supermarine's design proposal in response to the Air Ministry specification F7/30 for a replacement aircraft for the Bristol Bulldog, with which most of the squadrons in the area known as Air Defence of Great Britain were equipped. It was an all-metal-construction monoplane with a cranked wing and a fixed undercarriage complete with large undercarriage fairings and an open cockpit. Powered by the new Rolls-Royce Goshawk engine with evaporative cooling, it made its first flight in February 1934. Apart from its rather strange appearance, what was puzzling was the fixed undercarriage, especially since Supermarine had been devising raised undercarriages since 1920. Looking closely at the aircraft, one would have expected that the Type 224 would have had a drag-

A profile view of the Type 224.

168   The Schneider Trophy Air Races

The Type 224 in flight. Note the anhedral section of the wings and the fixed undercarriage covered in part by a fairing. It was no coincidence that the aircraft was known at the Supermarine works as Mitchell's Stuka.

The Polish PZL P.24 photographed at the Istanbul Aviation Museum.

reducing arrangement incorporated into the design. According to Supermarine, the cranked wing meant that the undercarriage would be much shorter and in addition a gun could be housed in the fairing without the need to disturb the leading edge of the wings. Such was the despair at the Air Ministry that there was even talk of purchasing the Polish PZL P.24 Fighter which was faster than the Type 224. This was all well and good, but despite a relatively powerful engine and satisfactory armament, the P.24 had a fixed undercarriage and was a gull-winged monoplane. Added to that, it could not stand up against some of the Axis fighters such as the Macchi MC.200 and the Messerschmitt Bf 109. Understandably, the Type 224 was not a success, and one is tempted to join with the Supermarine works team in calling the aircraft 'Mitchell's Stuka'; certainly the Junkers 87 Stuka looked remarkably similar! Apart from that, the rate of climb of the Type 224 was below specification, the cooling system was far from being perfected and the aircraft itself was overweight. Sir Robert McLean, the chairman of Vickers (Aviation) Ltd, believed that the failure of the Type 224 was largely down to the nature of the F7/30 specification, in particular the low landing speed which led to a large wing area. This was a basically unfair comment, but he drew some consolation from the news that rival companies had also failed to produce an aircraft to match the specification. In fact, there is some justification in the comment that the Type 224 did not emerge from the successful Schneider Trophy aircraft, but did incorporate much of the experience gained by Mitchell's team in the design and construction of racing seaplanes. The void created by the failure of F7/30 left a late entry by Gloster, who proposed a fabric-covered biplane with an obsolescent airframe and a fixed undercarriage powered by an air-cooled radial engine. Surprisingly Gloster won the contract with the Gloster SS.37, an aircraft that became known as the Gloster Gladiator. It was the last British biplane fighter to be manufactured and went into

The Gloster Gladiator Mark 1.

service in 1937 with 72 Squadron. With a top speed of 242 mph and climb rate of 15,000ft in six and a half minutes, it was the first military aircraft to feature an enclosed cockpit.

However, by mid-1934 Mitchell was already working on a radically-revised version of the Type 224 amid a steadily worsening international situation in Europe. By the time Mitchell had submitted his redesign of the Type 224 to the Air Ministry, Germany had withdrawn from the Disarmament Conference in Geneva and left the League of Nations. As a result of Germany's departure from Geneva, the British government began looking urgently at the worst deficiencies of the nation's defences and agreed almost unanimously that priority must be given to air defences. This required a further forty-one squadrons to be fully equipped by 31 March 1939 and was announced to Parliament on 19 July 1934. A week later, Mitchell's redesign of the Type 224 was sent to the Air Ministry.

Reginald Mitchell.

The new proposal removed the anhedral or downward angle section of the wings, reducing the span to 39ft 4in and the fixed landing gear was replaced by a retractable undercarriage. The armament of four machine guns, two located in the fuselage and two in the wings, represented an improvement, but the aircraft was still powered by the evaporative-cooled Goshawk engine and it was a long way from

The Bellanca 28-70 *Irish Swoop*. Mitchell may have been influenced by the aircraft's retractable undercarriage in his redesign of the Type 224. The photograph shows the Bellanca being weighed in one of Mildenhall's hangars days before the 1934 MacRobertson Air Race.

the killer fighter envisioned by the Air Ministry. An improvement on the previous design it may have been, but the official response from the Air Ministry was still lukewarm. It is interesting to note that two of the entries for the Macpherson-Robertson England to Australia Challenge of 1934, the rather sleek DH 88 Comet and the Bellanca 28-70 *Irish Swoop*, both had retractable undercarriages. There is no evidence to suggest that Mitchell was swayed by these designs, but he did inspect the Bellanca most carefully when it flew down to Southampton for the fitting of its cowling. However, the aircraft was withdrawn before the race as it had not completed official load testing for fuel capacity.

It was probably at this point that Sir Robert McLean made the decision to allow Mitchell and his design team to proceed with the specification without interference from the Air Ministry. In November 1934 the Vickers Board voted in favour of a resolution allowing Mitchell to continue with the design of the Type 300, as it was now known at Supermarine, along with £10,000 as a mark of the faith they placed in the designer of the Schneider Trophy aircraft. Sir Robert subsequently wrote to the Air Ministry informing them that Mitchell's design was proceeding at company expense and no official interference would be tolerated. From Mitchell's point of view, no design could proceed fully without official consultation and while he valued the support of the Vickers Board, he decided to steer a careful course to ensure that relations with the Air Staff and Farnborough continued to run smoothly:

> But the design of an aeroplane is a large venture, in which, one way and another, many people are involved. By the 1930s, and indeed much earlier, these people – in the firms, in the Air Ministry, at Farnborough and Martlesham Heath – had formed habits and friendships. It was not easy to break these habits and friendships by an administrative fiat. This was particularly so with Mitchell in 1934 and 1935 for he was leading a threatened life; he had already had an operation for cancer and while he hoped that this had been completely successful, he looked coolly at the alternative. If time were to be short he would need the best advice.

Martlesham Heath was home to the Aeroplane and Armament Experimental Establishment (A&AEE). The A&AEE carried the evaluation and testing of many of the aircraft types, including the Spitfire and Hurricane, and much of the armament and other equipment that would later be used during the Second World War. However, more immediately, time was indeed short. Mitchell had consulted his doctor as early as August 1933; he diagnosed rectal cancer and an operation was performed almost at once, after which he went to Bournemouth to convalesce. Yet unbeknown to Mitchell the prognosis was not good and the designer only had a maximum of four years left to live.

In the meantime Rolls-Royce had been working on a supercharged V-12 engine with a capacity of 27 litres and a target output of 1,000hp. In July 1934 a developmental model of this engine, known as the PV-12, completed its 100-hour bench test and although the engine could only muster 790hp, Rolls-Royce engineers were confident of an eventual output of 1,000hp. In December 1934 Mitchell made the decision to incorporate the new engine into the Type 224, an engine that eventually became the famous Merlin. Both the aircraft and the engine were now supported by private finance, indeed the 'PV' prefix stood for Private Venture, but notwithstanding Sir Robert's attitude to Air Ministry interference, work on the aircraft proceeded through 1934 with the Air Staff, in particular Air Marshal Sir Hugh Dowding (later to command RAF Fighter Command), taking an active interest in developments. In December 1934 it appeared that the period of short-lived private finance had come to an end as Hugh Dowding placed an order with Supermarine for a single prototype aircraft described in the F37/35 specification, which was virtually written around the information that had been presented by Supermarine. Thus, the F37/34 specification did *not* result in the prototype Spitfire: the reverse was actually true; the specification was written to describe the aircraft that Supermarine were already working on and was signed in January 1935. Rather an odd way for aircraft procurement to proceed, but the international situation was hardly conducive to orderly formality. As far as the name Spitfire was concerned it is said that when it came to giving the new single-seat fighter a name, McLean suggested Spitfire, the affectionate term he used for his spirited elder daughter. Initially the Air Ministry had reservations about calling the aircraft the Spitfire, as did Mitchell, who argued for calling the new aircraft the Shrew, but McLean's wishes prevailed. According to Morgan and Shacklady's book *Spitfire: The History*, the Air Ministry agreed, arguing that the name should begin with the letter 'S' and suggest something venomous, hence Spitfire.

The F37/34 requirement for a four-gunned aircraft was superseded in April 1935 by F10/35 which called for a greater lethality of armament in the form of six, preferably eight, machine guns in each wing. Squadron Leader Ralph Sorley was particularly associated with the notion that for a fighter to inflict lethal damage to an all-metal bomber it would be necessary for an attacking aircraft to achieve no fewer than 256 hits. At the same time, a fighter attacking a bomber flying at 180 mph would be most unlikely to fire more than a two-second burst. The arithmetic was simple. With each gun firing at a rate of 1,000 rounds per minute, at least 8 guns would be required to achieve the lethality required. Thus eight guns became the policy and Sorley had it written into the F10/35 specification. The additional

Sir Hugh Dowding.

Squadron Leader Ralph Sorley. The photograph was taken in his later life.

guns initially caused a headache for Mitchell's design team and represented a key departure from established practice in that the guns now had to be mounted in the wings and firing outside the propeller disc. There was now no way of clearing stoppages in the air, but at least any interrupter equipment would be eliminated and the rate of fire would be achieved:

> So when Ralph Sorley came along and wanted two more guns added to each wing it presented a considerable problem for Mitchell, whose wing was very much thinner than Camm's wing on the Hurricane. Mitchell was determined not to ease up on his thickness/chord requirements and his staff pointed out that with an elliptical plan form, a greater chord and hence a greater actual thickness could be maintained further out along the wing span than with a straight tapered wing. This would give just enough room to install two more guns each side outboard of the original four-gun installation.

Apart from the armament, several other important changes were incorporated into the design during the course of 1935. The stream cooling of the Goshawk had never been satisfactorily resolved and it was the result of a visit to America by Cyril Lovesey of Rolls-Royce that ethylene glycol cooling was adopted, resulting in a smaller radiator that fitted neatly into the recess that was planned for the previous cooling system in the starboard wing. Lovesey was a key figure in the development of the Rolls-Royce Merlin and was the company representative for support of the Rolls-Royce R engine during its trials at Calshot for the Schneider Trophy races in 1929 and 1931. Readers will recall that it was Lovesey who correctly estimated the petrol consumption in Waghorn's S.6 in 1929.

The second change was the move to the elliptical thinner wing that was to become the distinctive visual feature of the Spitfire. The practicalities that led to the change in wing shape included the fact that the new engine was heavier than the Goshawk's and the leading edge of the wing needed to be brought forward in order to accommodate it. It also had

Beverley Shenstone worked with Mitchell on the Spitfire's elliptical wing.

certain other attractions such as a reduced drag in level flight conditions at altitude, an explanation that is important as it had been fairly widely suggested at the time that the elliptical wing of the Spitfire was a copy of the Heinkel HE.70 transport, which had been used by Rolls-Royce in 1936 as a test-bed for the Kestrel engine. While the Supermarine team had undoubtedly studied some of the excellent features of the HE.70, the elliptical wing shape was not a feature they copied. In fact the S.4 at Baltimore in 1925 had an elliptical wing, as did the Italian Piaggio-Pegna P.c.7 that was prepared for the 1929 Schneider Trophy race. Another consideration was that the wing shape of a transport aircraft such as the HE.70 that first flew in 1932 was an implausible model for a new breed of fighter whose wing shape was particularly important when it came to climb and speed. There was a postscript to this story which involved Beverley Shenstone, the Canadian aerodynamicist on Mitchell's design team. Not only did he debunk the Heinkel theory in an interview with the late Alfred Price, but went on to credit the German company with one aspect of the design:

> It has been suggested that we at Supermarine had cribbed the elliptical wing shape from that of the Heinkel 70 transport. That was not so. The elliptical wing shape had been used on other aircraft and its advantages were well-known. Our wing was much thinner than that of the Heinkel and had a quite

different section. In any case it would have been simply asking for trouble to have copied a wing shape from an aircraft designed for an entirely different purpose. The Heinkel 70 did have an influence on the Spitfire, but in a rather different way. I had seen the German aircraft at the Paris Aero Show and had been greatly impressed by the smoothness of its skin. There was not a rivet head to be seen.... I was so impressed that I wrote to Ernst Heinkel, without Mitchell's knowledge, and asked how he had done it, was the aircraft's skin made of metal or wood? I received a very nice letter back from the German firm, saying the skinning was of metal with the rivets countersunk and very carefully filled before the application of several layers of paint.

Armed with this information, Shenstone incorporated the sunken rivets, as outlined by Ernst Heinkel, into the Spitfire's elliptical wing.

The prototype Spitfire was built at the Supermarine works at Woolston and transported by lorry to Eastleigh where it was put back together under the supervision of Ken Scales in preparation for its first test flight by Joseph 'Mutt' Summers, the chief test pilot of Vickers (Aviation) Ltd. The prototype was still in its unpainted format with the works protective treatment on its metal surfaces and fitted with a fine-pitch wooden propeller in order to give the aircraft higher revs on take-off and to minimize the effects of torque. This must have been some relief to both Mitchell and Summers as it ensured a safe take-off run. In fact the Spitfires used a variety of propellers during their time in service and, with the advent of the Merlin III engine in 1938, the two-bladed wooden propeller was replaced by the de Havilland three-blade metal two-pitch propeller, significantly enhancing performance, particularly in the climb.

Quill watched the prototype taxi towards one of the four large lights that were situated around the perimeter and turn into wind:

> The aeroplane was airborne after a very short run and climbed away comfortably. Mutt did not retract the undercarriage on that first flight – deliberately of course – but cruised fairly gently around for some minutes, checked the lowering of the flaps and the slow flying and stalling characteristics, and then brought K5054 in to land. Although he had less room than he would have liked, he put the aeroplane down on three points without too much 'float', in which he was certainly aided by the fine-pitch setting of the propeller.

When the aircraft finally came to a standstill and the engine was shut down, everyone crowded round Summers to hear the verdict. As he pulled off his helmet and clambered out of the machine he said firmly: 'I don't want anything touched.'

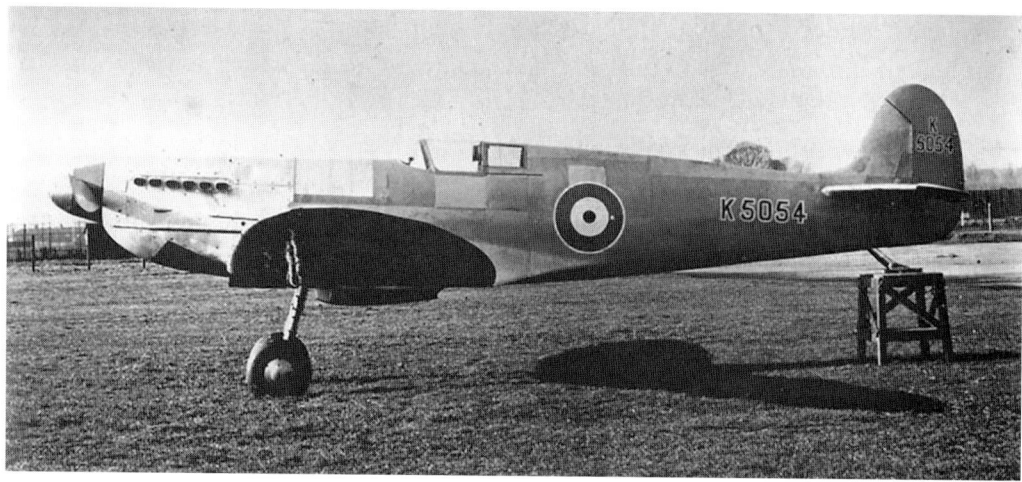

The prototype Spitfire prior to its maiden flight. Note the two-bladed propeller and the flat hood covering the cockpit.

It was a remark that was to become widely misinterpreted in Spitfire folklore, the implication being that the aeroplane was perfect in every way, which was a ridiculous and unrealistic idea. After the initial test flight of some fifteen minutes, during which one take-off and one landing had been accomplished, Summers was

Mitchell sitting on the running board of his car after the first flight of the Spitfire at Eastleigh on 5 March 1936. He is surrounded by (left to right): Mutt Summers, Anthony Payne of Supermarine, S. Scott Hall of the Royal Aeronautical Society and Jeffrey Quill.

far too experienced to make any such sweeping statement and simply meant that there were no major snags noted before his next flight. However, it was clear that the first flight had been very successful and as Quill flew a smiling Summers back to Brooklands in the firm's Miles Falcon Six, Summers was obviously very pleased. After they had put the Falcon away and were in the famous Brooklands Flying Club bar, the prototype Hurricane was standing in the next hangar, prompting Quill to write:

> So the two new fighter aircraft – destined four years later to save our country in time of war – had now both flown in prototype form. Neither was anywhere near being a practical fighting machine, nor was either yet ordered in quantity by the Royal Air Force, so much work still remained to be done.

Nevertheless, the question remained, how would the aircraft perform under the stringent test conditions at Martlesham compared to Sydney Camm's Hurricane and would the testing result in a production order from the Royal Air Force? However, as Quill wrote, there remained a great deal of work to do, not least on the maximum speed of the aeroplane at altitude. Quill on one of his early flights noted with some disappointment that the aeroplane would only achieve a maximum speed of 335 mph at less than 17,000ft and Mitchell had his heart set on a speed of 350 mph. Quill and Summers knew that Mitchell would not let the Spitfire go to Martlesham until it did, particularly as rumour had it that the Hurricane was already at Martlesham and was showing a speed of about 325 mph. Therein lay the difficulty: would the RAF choose the largely fabric-covered Hurricane which was much cheaper to produce, or the all-metal Spitfire, which at the time was only performing slightly better and was more expensive to produce. Clearly the Air Ministry was looking at Mitchell, with the experience of the Schneider Trophy racers behind him, to produce an aircraft that offered a substantial speed difference over the Hurricane. It was true to say that Mitchell was a worried man as he sought to find the cause of the disheartening speed of the Spitfire. Eventually suspicion fell on the fixed-pitch wooden propeller, and after a trial flight on 27 March with a new propeller, this time with modified tip sections, the aircraft achieved a speed of 348 mph and K5054 was flown to Martlesham on 26 May by Mutt Summers.

Flight Lieutenant Humphrey Edwardes-Jones, known as EJ to his friends, commanded A Flight at the A&AEE at Martlesham and was normally an individual who was almost entirely unflappable, but he does confess to being a little flabbergasted by his orders on 26 May 1936. He was told to fly the Spitfire immediately it landed at Martlesham and thereafter telephone Air Marshal Sir Wilfred Freeman at the Air Ministry. No reason was given for these strange orders. Accordingly, Edwardes-Jones refuelled the aeroplane and, after a cockpit briefing, took off on a short flight around the local area and landed again. His resulting conversation with Sir Wilfred Freeman, which did not mention that he had almost forgotten to lower the undercarriage, was short and to the point: was the Spitfire, in EJ's opinion, capable of being flown safely by the ordinary service-trained fighter pilot, more particularly those emerging from the training sequence being set up to meet the new expansion programmes? Fortunately, EJ answered with an unequivocal yes. His positive reply about pilots emerging from the current training system, providing they were properly instructed in the use of retractable undercarriage, flaps and other systems associated with the new aircraft coming into service with the RAF was exactly what Freeman wanted to hear. EJ went on to say that pilots would have no difficulty with the Spitfire and it was a delight to fly. Thus EJ became the first RAF pilot to fly the Spitfire and one week later on 3 June the production order for 310 Spitfires was signed. By the time of the Munich Conference in September 1938, the Spitfire was in production.

On 11 June 1937 Reginald Mitchell died at home after a visit to see Doctor Anton Lowe in Vienna proved fruitless. He was only 42 years old at his death but continued to work despite increasing pain, tweaking the design of the Spitfire up to

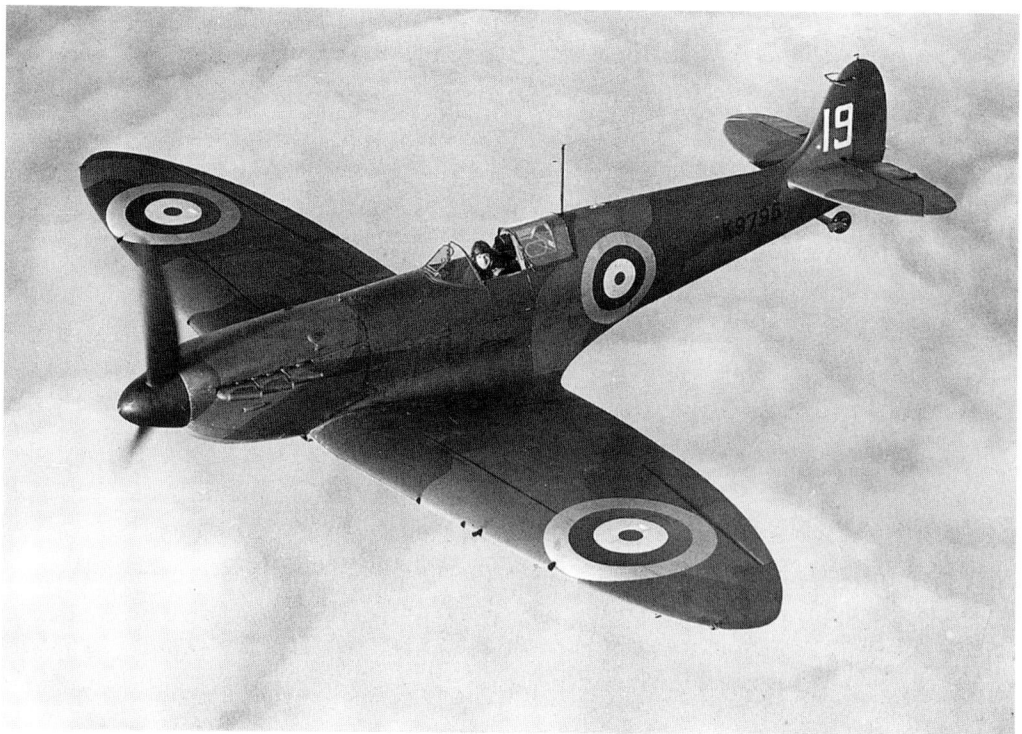

The Spitfire Mark 1A flying with 19 Squadron. The flat hood indicates that the photograph was taken prior to 1939.

the moment of his demise. After Mitchell's death Joseph Smith took over as chief designer; a practical man who understood the Spitfire in its entirety having joined Supermarine in 1921. He had been working under Mitchell's wing since 1926 and led the team that developed the Spitfire and its successors into the transonic age. It was he that modified the Spitfire canopy with a new form of blown canopy in 1939 following suggestions from pilots. This improved headroom and enabled better vision laterally and to the rear. At the same time the manual hand-pump for operating the undercarriage was replaced with a hydraulic system driven by a pump mounted in the engine bay.

The version of the Spitfire that fought in the Battle of Britain was powered by the Merlin II engine of 1,030hp and reached a maximum speed of approximately 360 mph with a ceiling of 34,000ft. Faster than its formidable German opponent, the Messerschmitt Bf 109, at altitudes above 15,000ft, Spitfires were sent by preference to engage German fighters while the slower Hurricanes went for the bombers. More Hurricanes than Spitfires served in the Battle of Britain and they were credited with more kills, but arguably the Spitfire's superior high-altitude performance provided the margin of victory. Although they were different aircraft, they complemented each other perfectly and it took both machines to win the battle that raged above southern England.

One of the most important modifications to the Spitfire was replacing the machine-gun armament with wing-mounted Hispano 20mm cannon. In December 1938, Joseph Smith was instructed to design a Spitfire with a single Hispano mounted under each wing. Smith objected to the idea and designed an installation in which the cannon were mounted on their sides within the wing, with only small external blisters on the upper and lower wing surfaces covering the sixty-round magazine. The first Spitfire armed with a single Hispano in each wing was L1007 and was posted to RAF Drem in Scotland in January 1940 for squadron trials. On 13 January L1007, armed with the new Hispano cannon and piloted by Pilot Officer Proudman of 602 Squadron, took part in an engagement when a Heinkel He 111 was shot down. Soon after this Supermarine was contracted to convert thirty Spitfires to take the cannon-armed wing and 19 Squadron received the first of these in June 1940. The first series production Spitfires to enter squadron service with a 20mm Hispano cannon and two machine guns in each wing was the Mark VB which entered service in 1941. A further modification that was to save the lives of many pilots took place in late in 1940: a Martin-Baker-designed quick-release canopy mechanism began to be fitted to all Spitfires. The system employed unlocking pins, actuated by cables operated by the pilot pulling a small red rubber ball mounted on the canopy arch. When freed, the canopy was taken away by the slipstream.

The last word on the Spitfire quite fittingly comes from a 20-year-old pilot who flew the Spitfire up until the end of the war when 74 Squadron was called home to convert to the Meteor jet fighter. Handing over their Mark XVI Spitfires to 485 Squadron, he wrote in his diary that he 'would certainly miss the incomparable Spitfire, legendary and lethal, yet so graceful and unforgettable'. Flying back over the Channel to England, he was reminded of the day he had first gone solo in a Spitfire in 1943, writing that it was 'a day he would never forget'. That pilot was my father.

Appendix

Schneider Trophy Results
(Pilots and aircraft that completed the course)

First Contest: 16 April 1913 at Monaco. Seven entries with two completing the course:

Aircraft	Engine	Pilot	Country	Speed
Deperdussin	160hp Gnome	Maurice Prévost	France	45.75 mph
Morane-Saulnier	80hp Gnome	Roland Garros	France	30.42 mph

Second Contest: 20 April 1914 at Monaco. Twelve entries with two completing the course:

Sopwith Tabloid	100hp Gnome mono	Howard Pixton	Britain	86.78 mph
FBA flying boat	100hp Gnome mono	Ernst Burri	Switzerland	51.00 mph

Third Contest: 10 September 1919 at Bournemouth. Seven entries, not one completed the course:

Savoia S.13	250hp Isotta Fraschini	Guido Janello	Italy	

Janello failed to complete the eleven-lap course satisfactorily and was disqualified. The next contest was awarded to Italy by way of recompense.

Fourth Contest: 21 September 1920 at Venice. Four entries with one completing the course. An Italian fly-over by Luigi Bologna:

Savoia S.12	550hp Ansaldo	Luigi Bologna	Italy	107.22 mph

Fifth Contest: 7 August 1921 at Venice. Four entries with one completing the course:

Macchi M.7	250hp Isotta Fraschini	Giovanni de Briganti	Italy	117.90 mph

Sixth Contest: 12 August 1922 at Naples. Six entries with four completing the course:

Sea Lion II	450hp Napier-Lion	Henry Biard	Britain	145.7 mph
Savoia S.51	300hp Hispano-Suiza	Alessandro Passeleva	Italy	143.20 mph
Macchi M.17	260hp Isotta Fraschini	Arturo Zanetti	Italy	139.75 mph
Macchi M.7	260hp Isotta Fraschini	Piero Corgnolino	Italy	123.7 mph

Seventh Contest: 28 September 1923 at Cowes. Nine entries with three completing the course:

Aircraft	Engine	Pilot	Country	Speed
Curtiss CR-3	465hp Curtiss D-12	David Rittenhouse	America	177.38 mph
Curtiss CR-3	465hp Curtiss D-12	Rutledge Irvine	America	173.46 mph
Sea Lion III	550hp Napier-Lion	Henry Biard	Britain	157.17 mph

Eighth Contest: 26 October 1925 at Baltimore. Eight entries with three completing the course:

Aircraft	Engine	Pilot	Country	Speed
Curtiss R3C-2	600hp Curtiss V-1400	James Doolittle	America	232.57 mph
Gloster III	700hp Napier-Lion	Hubert Broad	Britain	199.17 mph
Macchi M.33	435hp Curtiss D-12	Giovanni de Briganti	Italy	168.44 mph

Ninth Contest: 13 November 1926 at Hampton Roads. Seven entries with four completing the course:

Aircraft	Engine	Pilot	Country	Speed
Macchi M39	882hp Fiat AS.2	Mario de Bernardi	Italy	246.50 mph
Curtiss R3C-3	600hp Curtiss V-1400	Frank Schilt	America	231.36 mph
Macchi M.39	882hp Fiat AS.2	Adriano Bacula	Italy	218.01 mph
Curtiss F6C-1 Hawk	440hp Curtiss D-12	William Tomlinson	America	136.95 mph

Tenth Contest: 26 September 1927 at Venice. Seven entries with two completing the course:

Aircraft	Engine	Pilot	Country	Speed
Supermarine S.5	900hp Napier-Lion VIIA	Sidney Webster	Britain	281.66 mph
Supermarine S.5	875hp Napier-Lion VIIB	Oswald Worsley	Britain	273.01 mph

Eleventh Contest: 7 September 1929 at Calshot. Six entries with three completing the course:

Aircraft	Engine	Pilot	Country	Speed
Supermarine S.6	1,900hp Rolls-Royce R	Richard Waghorn	Britain	329.63 mph
Macchi M.52R	1,000hp Fiat AS.3	Tommaso Dal Molin	Italy	284.20 mph
Supermarine S.5	875hp Napier-Lion VIIB	D'Arcy Greig	Britain	282.11 mph

Final Contest: 13 September 1931 at Calshot. Three British entries with one completing the course. A British fly-over by John Boothman:

Aircraft	Engine	Pilot	Country	Speed
Supermarine S.6B	2,350hp Rolls-Royce R	John Boothman	Britain	340.08 mph

Select Bibliography

Primary Sources:
Almond, Peter, *A Century of Flight* (W.H. Smith, 2002)
Babington Smith, Constance, *Testing Time* (Harper & Brothers, 1961)
Barker, Ralph, *The Schneider Trophy Races* (Chatto & Windus, 1971)
Biard, Henry, *Wings* (Hurst & Blackett, 1934)
Cowin, Hugh, *Allied Aviation of World War 1* (Osprey, 2000)
Gibbs-Smith, Charles, *Aviation* (Science Museum, 2003)
Hawkes, Ellison, *The Schneider Trophy Contests (1913–1931)* (Real Photographs, 1945)
Mondey, David, *The Schneider Trophy* (Robert Hale, 1975)
Morgan, Eric and Shacklady, Edward, *Spitfire: The History* (Key Publishing, 2000)
Nicolaou, Stéphane, *Flying Boats and Seaplanes* (MBI, 1998)
Pixton, Stella, *Howard Pixton: Test Pilot and Pioneer Aviator* (Pen & Sword, 2014)
Price, Alfred, *The Spitfire Story* (Silverdale, 2002)
Quill, Jeffrey, *Birth of a Legend: The Spitfire* (Quiller, 1986)
Robertson, Bruce, *Sopwith: The Man and His Aircraft* (Harleyford, 1970)
Shelton, John, *Schneider Trophy to Spitfire* (Haynes Publishing, 2008)
Simmonds, Graham, *Early French Aviation* (Air World, 2019)
Treadwell, Terry, *British & Allied Aircraft Manufacturers* (Amberley, 2011)
Vorderman, Don, *The Great Air Races* (Bantam Books, 1991)
Webb, Denis, *Never a Dull Moment* (J. & K. Publishing, 2001)
Wilkins, Mark, *German Fighter Aircraft in World War 1* (Casemate, 2019)

Secondary Sources:
'My First Ten Years in Aviation' by Sir Thomas Sopwith in the *Journal of the Royal Aeronautical Association*, April 1961
'R.J. Mitchell, Aircraft Designer', by Joe Smith in *The Aeroplane*, January 1954
'The Schneider Trophy 1929' by Flight Lieutenant H.R.D. Waghorn in the *Journal of the Royal Aeronautical Society*, May 1930

Index

Acosta, Bertrand, 70, 71, 77–8, 107
Aerial Derby, 22, 23, 31, 32, 35, 60, 61, 66, 101, 102
Alcock, Capt John, 66–7, 82
American National Aeronautic Association, 110, 119
Antoinette (aircraft), xii, 2, 3
Atcherley, F/O Richard, 139, 140, 151, 152, 161

Bacula, Lt Adriano, 120, 125, 126
Balbo, Gen Italo, 83, 84, 85, 144, 145, 147, 150, 154, 157
Battle of Britain, xiv, 61, 179
Bellanca, 28-70, 170, 171
Benoît, Jean, 24
Berthold, *Hauptmann* Rudolf, 45, 46
Bettis, Lt Cyrus, 122, 123
Biard, Henry, 94–5, 96, 98, 99, 102, 107, 108, 111, 112–13, 115, 116, 118
Bider, Oskar, 24, 25, 26
Blériot (aircraft), xii, 1, 8, 11, 12, 17, 19, 21, 22, 23, 24, 25, 26, 49, 57, 60, 91
Blériot, Louis, xii, 1, 3, 4, 8, 46, 68, 76
Boelcke, *Hauptmann* Oswald, 45, 46
Bologna, Lt Luigi, 93
Boothman, F/Lt John, xiv, xx, 20, 161, 162, 164, 165
Brindejonc de Mouliais, Marcel, 30, 33, 52
Brinton, Lt Gerald, 161, 162
Bristol M1 Monoplane, 47, 48
Broad, Hubert, 109–10, 111, 113, 116, 117, 118
Brooklands, 19, 20, 23, 35, 36, 53, 63, 177
Burri, Ernest, 34, 39, 41, 43
Byrd, Com Richard, 71–2, 75, 77

CAMS (aircraft), 92, 96, 103, 104, 107
Carberry, Lord John, 35, 39, 40, 62–3
Casale, Lt Jean, 91
Castoldi, Mario, xix, xxi, xxii, 115, 119–20, 128, 132, 145, 166

Centurione, Vittorio, 120, 121, 123, 126
Coanda (aircraft), 46, 49
Coandă, Henri, 45, 47
Cobham, Alan, 72–3, 86
Cockburn, George, 3, 4, 9, 58, 59
Cody, Samuel, 19, 20, 21, 30–1, 49
Conant, Lt Hersey, 122, 123
Conneau, Lt Jean (André Beaumont), 15, 17, 18, 19–20, 53
Coppa Schneider, xxii, 93
Corbett-Wilson, Denys, 59–60
Cuddihy, Lt George, 111, 123, 125, 126
Curtiss (aircraft), xiv, 50, 70, 71, 82, 83, 84, 90, 100, 104, 105, 106, 107, 108, 109, 110, 111, 113, 114, 115, 117, 118, 120, 121–3, 124–5, 126, 130, 132, 133
Curtiss, Glenn, 3, 4–5, 14

Dal Molin, W/O Tommaso, 150, 151, 153, 158
de Bernardi, Maj Mario, xix, 120–1, 123, 125, 126, 127, 129, 134, 137, 138
de Brigante, Giovanni, 93, 115, 117, 118
de Havilland (aircraft), 57, 43, 77, 78
de Havilland, Geoffrey, 48, 57
Delagrange, Léon, 5, 12–13
Del Prete, Carlo, 80, 84
Deperdussin, Armand, 29–30
Deperdussin (aircraft), 20, 22, 25, 27, 28, 29, 31, 33, 34, 39, 40, 49, 60, 92
DH 88 Comet, 171
Dole, James, 76, 77, 79
Doolittle, Lt James, xvii, xix, 115, 117, 121, 123
Dowding, AM Sir Hugh, 172, 173

Earhart, Amelia, 78, 79
Edwardes-Jones (EJ), F/Lt Humphrey, 177–8

Fabre, Henri, 12, 13, 14
Farman (aircraft), 5, 6, 8, 9, 10, 23, 53, 56, 59, 61, 68, 69, 91

Index

Farman, Henri, 2, 3, 5, 9, 10, 24, 68
Farman, Maurice, 9
Ferrarin, Capt Arturo, 80–1, 120–1, 124, 125, 129, 134
Flying Flirt, the, xiii
Fokker, Anthony, 34–5, 51
Fokker (aircraft), 34, 46, 47, 51, 52, 69, 71, 77, 78, 79–80
Freeman, AM Sir Wilfred, 177, 178
Frey, André, 15, 17

Garros, Roland, 17, 18, 27, 28, 29, 31, 33, 40, 43, 51, 54
Gaudart, Louis, 26
Gilbert, Eugene, 15, 55, 62
Gloster II, xviii, 109–10, 111
Gloster III, 113, 114, 116, 117
Gloster Gladiator, xxii, 169–70
Goodfellow, Maj Alan, 152
Gordon Bennett Races, xi, xii, xiii, 3, 4, 11, 21, 22, 28, 31, 33, 56, 58, 59, 88, 91, 94, 128, 160
Gorton, Lt Adolphus, 106, 110
Grade, Hans, 6
Grahame-White, Claude, 8–9, 10, 11, 18, 56
Greig, F/Lt D'Arcy, 138, 139, 140, 144, 151, 153, 161

Hamble, river, 36, 37, 38
Hamel, Gustav, 21, 22, 30, 32
Hawker, Harry, 31, 32, 35, 36, 65, 66, 89
High Speed Flight (RAF), 18, 127, 128, 131, 135, 138, 139, 140, 156, 161, 165, 166
Hinkler, S/Ldr Bert, 114, 116–17
Hope, F/Lt Eustace, 161, 163
Houston, Lady Lucy, xx, 156, 157, 159, 165, 166

Irvine, Lt Rutledge, 105, 107, 108

Janello, Sgt Guido, 91, 92
Janoir, Louis, 25, 33, 39
Johnson, Amy, 79, 80

Kenworthy Lt Com Reginald, 101, 102
Kinkead, F/Lt Samuel, 128, 134, 137, 138
Köhl, Hermann, 78
König, Benno, 6, 7, 8

Laffan's Plain, 31
Latham, Hubert, xi, xii, 2, 3, 4

Lefebvre, Eugène, 3, 4, 5
Levasseur, Pierre, 14, 33, 39, 41, 43
Lindberg, Charles, 65, 75–6, 77, 87
Longton, F/Lt Walter, 100
Lovesey, Cyril, 151, 153, 173–4

Macchi, xix, xxii, 81, 93, 94, 97, 98, 113, 114, 118, 119, 120, 123, 126, 128, 129, 132, 133, 135, 137, 138, 144, 145, 147, 149, 151, 154, 157, 158, 166, 169
Mackenzie-Grieve, Lt Com Kenneth, 65, 66
Mahl, Victor, 37, 38, 42
Martlesham Heath (A&AEE), 140, 167, 171, 177
McClean, Frank, 61, 62
McLean, Sir Robert, 169, 171, 172
Mitchell, Gen Billy, 82–3, 104, 105
Mitchell, Reginald xiv, xv, xix, xxi, xxii, 50, 64, 86, 90, 94, 95, 98, 99, 102, 108, 111, 112, 115, 116, 118, 127, 128, 131, 132, 135, 137, 138, 139, 140, 142, 145, 146, 148, 156, 157, 161, 165, 166, 169, 170, 171, 172, 173, 175, 176, 177, 178
Military Aeroplane Competition, 48–9
Mollison, Jim, 77, 78, 79
Monaco, xvi, 13, 23, 25, 26, 27, 29, 30, 32, 33, 34, 36, 37, 39, 40, 41, 44, 50
Moore-Brabazon, John, 11, 57, 58
Morane (aircraft), 17, 18, 20, 21, 27, 28, 29, 30, 32, 33, 35, 39, 47, 51, 52, 55, 60
Mussolini, Benito, xix, xx, 83, 85, 109, 119, 126, 128, 144

National Air Tours, 127, 128
Nesterov, Lt Pyotr, 31, 32
Nicholl, Lt Col Vincent, 89
Nieuport (aircraft), 19, 20, 27, 28–9, 31, 32, 34, 39, 40, 46, 66, 91, 148, 149, 159, 160
Nieuport, Édouard, 21
Norton, Lt Harmon, 122, 123
Nungesser, Charles, 76

Ofstie, Lt Ralph, 115, 116, 117
Ogilvie, Alec, 21, 58–9
Orlebar, S/Ldr Augustus, 139, 140, 142, 151, 153, 161, 163, 164
Orteig, Raymond, 65, 75

Paris to Rome Race, 17
Passeleva, Alessandro, 96, 98, 99, 128

Paulhan, Louis, 9–11, 23, 53, 54
Pinedo, Francesco de, 84
Pixton, Howard, xvi, 19, 31, 33, 34, 35, 36–8, 40–2
Prévost, Maurice, xiv, 17, 25, 29, 33, 91
Price, Lt John, 107

Quill, Jeffrey, 167, 175, 176, 177
Quimby, Harriet, 22–3

Reims Air Meeting, 1, 9, 12, 14, 21, 58, 59
Rhodes-Moorhouse, 2/Lt William, 22, 23, 61
Rickenbacker, Capt Edward, 70
Rittenhouse, Lt David, 107, 108
Roe, Alliott Verdon (AV Roe), 63
Rolls, Charles, 11, 12, 57
Rolls-Royce, engines, xx, xxii, 139, 142, 146, 151, 156, 157, 161, 163, 167, 172, 173–4
Round Britain Race, 18–20, 31
Round Britain Seaplane Race, 31
Round Europe Race, 17–18
Royal Aero Club, 11, 18, 48, 57, 63, 79, 88, 92, 100, 110, 118, 119, 139, 151, 152, 156, 157, 158, 159
Royal Flying Corps, 31, 45, 46, 47, 49, 57, 58, 60, 61, 63, 119, 140
Royal Naval Air Service, 44, 45, 46, 57, 58, 59, 61, 62–3, 95, 114, 137
Royce, Henry, xix, 139
Rumpler-Taube, 6, 7

Sadi-Lecointe, Lt Joseph, 90, 91, 93, 148
Schilt, Lt Frank, 123, 125, 126
Schneider, Jacques, xi, xiii, xiv, xv, xvi, xviii, xxi, 14, 42, 85, 138, 139
Schofield, F/O Harry, 128, 134
Schwann, Commander Oliver, 15
Scott-Paine, Hubert, 50, 94, 95, 98, 99
Shenstone, Beverley, 174–5
Short (aircraft), 50, 61, 85, 86, 87, 130
Sigrist, Fred, 36
Sippe, Lt Sydney, 15, 62
Smith, Joseph, 179
Smith, Keith, 67
Smith, Ross, 67, 68
Snaith, F/O Leonard, 161, 165

Sopwith (aircraft), xx, xxi, 19, 31, 32, 35, 36, 38, 40, 45, 46, 59, 63-4, 66, 89, 90, 100, 140
Sopwith, Tommy, 23, 32, 35–6, 37, 38, 39, 41, 44, 63
Sorley, S/Ldr Ralph, 172, 173
Stainforth, F/Lt George, xx, 139, 140, 144, 161, 162, 164, 165–6
Stoeffler, Ernst, 34, 39
Supermarine Sea Lion I, 89–90
Supermarine Sea Lion II, 94–6, 97, 98, 99, 102
Supermarine Sea Lion III, xvi, 102, 107, 111
Supermarine Spitfire, xvi, xxv, 1, 111, 138, 148, 167, 171, 172, 174, 175, 176, 177–8, 179, 180
Summers, Joseph, 175, 176–7

Taylor, Charlie, x
Teste, Paul, 104
Thaw, William, 34
Tomlinson, Lt William, 123, 124, 125, 126
Train, Louis, 15, 16
Trenchard, MRAF Sir Hugh, 35, 59, 155, 157
Type 224, 167, 168, 169, 170, 172

Valentine, James, 17, 20, 60
Védrines, Jules, 15, 17, 18, 19, 20, 22, 54, 55
Venice, xiv, 19, 53, 93, 96, 127, 132, 133, 134, 135, 153
Voisin (aircraft), 9, 12, 13, 23, 24
Voisin, Gabriel, 12, 24

Waghorn, F/Lt Richard, xx, 139, 140, 142, 143, 151, 153–4, 155, 161, 174
Wead, Lt Frank, 106, 107
Webster, F/Lt Sidney, xix, 128, 130, 134, 135, 137, 140, 153
Weymann, Charles, 17, 19, 21, 26, 28, 29, 34, 40, 43
Williams, Lt Alford, 109, 132, 133, 149
(Witten) Brown, Arthur, 66, 67
Woolston, 94, 142, 175
Worsley, F/Lt Oswald, 128, 131, 134, 137
Wright Brothers, x, xi, xiii, 59, 66
Wright Flyer, x, xi, xii, 12

Zanetti, Arturo, 97, 98, 99